£17.50

INTRODUCTION TO INVERTEBRATE
CONSERVATION BIOLOGY

INTRODUCTION TO INVERTEBRATE CONSERVATION BIOLOGY

T. R. NEW

School of Zoology, La Trobe University
Bundoora, Victoria, Australia 3083

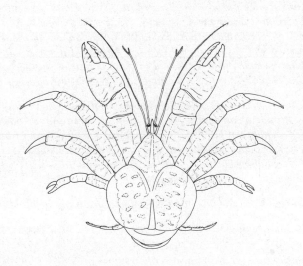

Oxford   New York   Melbourne
OXFORD UNIVERSITY PRESS
1995

Oxford University Press, Walton Street, Oxford OX2 6DP

Oxford   New York
Athens   Auckland   Bangkok   Bombay
Calcutta   Cape Town   Dar es Salaam   Delhi
Florence   Hong Kong   Istanbul   Karachi
Kuala Lumpur   Madras   Madrid   Melbourne
Mexico City   Nairobi   Paris   Singapore
Taipei   Tokyo   Toronto
and associated companies in
Berlin   Ibadan

Oxford is a trade mark of Oxford University Press

Published in the United States
by Oxford University Press Inc., New York

A catalogue record for this book is available from the British Library

Library of Congress Cataloging in Publication Data
New, T. R.
An introduction to invertebrate conservation biology / T. R. New.
Includes bibliographical references and index.
1. Invertebrates. 2. Wildlife conservation. I. Title.
QL362.N4   1995   33.95'516–dc201   94–45538

ISBN 0 19 8540523 (Hbk)
ISBN 0 19 8540515 (Pbk)

Typeset by the Electronic Book Factory Ltd, Fife, Scotland
Printed in Great Britain by
The Bath Press, Avon

PREFACE

Invertebrates are the most diverse and abundant animals in most natural ecosystems, but their importance in sustaining those systems is commonly not appreciated. This book is about the necessity and practice of conserving invertebrates to safeguard this predominant part of Earth's biotic heritage. Sound conservation management depends increasingly on sound knowledge of the biology and dynamics of the species and systems involved, so that much of the book emphasizes the relevance of understanding the biology of invertebrates and how to apply that understanding in conservation. My aim is to introduce the emerging science of invertebrate conservation to students, naturalists, biologists, and conservation managers, and to provide sufficient examples and references to indicate the progress that has been made and the directions in which work is proceeding in many parts of the world. Much of the information has hitherto been scattered in a wide range of publications or reports.

One important bias in coverage needs to be emphasized at the outset. Appreciation of the need for conserving invertebrates has developed almost entirely from initiatives from the more affluent temperate-region countries, and those experiences inevitably constitute the bulk of a summary such as this. This rather restricted geographical scope, which becomes especially evident in the lacunae among cases selected for appraisal in Chapter 7, reflects the massive political and economic diversity of the world's nations and its implications in differing commitment to practical conservation of obscure organisms. It does *not* imply any lack of need for invertebrate conservation in the tropics—indeed, some of the highest priority needs to assess and conserve biodiversity, and the greatest threats to its sustainability, occur in regions where invertebrate conservation as yet has no substantial role on conservation agendas. I hope that the book will complement the information accessible more easily in other texts on conservation and help readers to obtain greater appreciation of the vital roles of the animals that E. O. Wilson has elegantly termed 'the little things that run the world' (Wilson 1987). The underlying aims of biological conservation are now widely understood. Essentially, these are to understand and control human impacts on the natural world sufficiently to ensure that its other inhabitants may still be assured of a place to live and thrive, rather than be driven to extinction by destruction of their habitats and other vital needs. The *World conservation strategy* in 1980 (IUCN, UNEP, WWF 1980) proposed three specific objectives for the conservation of living species, namely:

1. To maintain essential ecological processes and life support systems on which human survival and development depend.

2. To preserve genetic diversity (the range of genetic material found in the world's organisms), on which depend the breeding programmes serving for the protection and improvement of cultivated plants and domesticated animals, as well as much scientific advance, technical innovation, and the security of many industries that use living resources.

3. To ensure the sustainable utilization of species and ecosystems (notably fish and other wildlife, forests and grazing lands) which support rural communities as well as major industries.

These themes have pervaded much recent discussion (IUCN, UNEP, WWF 1980), leading to the production of the sequel document *Caring for the world* (IUCN, UNEP, WWF 1991), and emphasize the need to maintain Earth's capacity to support human need, with concomitant maintenance of the great variety of life forms present.

However, many people equate 'conservation' simply with the 'saving' of notable species, rather than with broad and less tangible aims such as maintaining natural communities or whole ecosystems. The species are usually mammals or birds, animals with considerable public appeal and to which varying degrees of anthropomorphism can be transferred. Despite this widespread sympathy for vertebrates, which will surely continue to play a vital role in publicizing conservation issues widely, the well-being of most natural ecosystems may depend, rather, on the myriads of less-heralded animals, most of them invertebrates, whose collective ecological roles and influences predominate in the diverse communities they largely constitute. Appreciation of the values of invertebrates is increasing, and it is now clear that many need skilled attention to ensure that they, and the systems in which they thrive, continue to exist. This emerging attention, both to notable invertebrate taxa forming the focus of significant conservation activity, and more generally to their use as sensitive indicators and monitors of environmental health, is the theme of this short book. It provides an outline of the place and future of invertebrates in conservation biology, demonstrates how they enrich the scope and capabilities of practical conservation, and discusses many of the ideas and constraints that are apparent.

Practical conservation is, in one sense, a 'time-limited science'. Rates of change, and intensity of human impacts on the natural world, are severe and extensive, and have led to many extinctions already. Many people estimate that millions of species will become extinct in the next few decades. One weakness in political argument for conservation is the lack of hard data needed to make the best possible decisions to counter this alarming forecast, coupled with the impracticality of ever obtaining it in time to do so. This knowledge impediment is especially severe for invertebrates, as will become evident in the early chapters. It is not possible, for instance, to name or recognize all the species of most larger taxonomic groups prevalent in any community or assemblage. The main needs are

to seek and adopt methods for rapid detection and assessment of biota, to rank the species or faunas in some logical way(s) for conservation importance or priority, and to implement management and practical conservation from incomplete data. As the most numerous components of organismal diversity, the invertebrates are of paramount importance in pursuing such advances.

Each chapter concludes with a few suggestions for allied reading, and publications cited in the main 'References' section generally extend to mid-1993, although a few more recent ones are included. Many are cited as examples only, and themselves have substantial additional bibliography. I have opted to include a substantial number of references, though I appreciate that many practitioners of conservation will not have ready access to major institutional libraries. I try to give sufficient details of the work under discussion to make the book reasonably self-contained. For those wanting more information, the references will serve as an entry to the primary and secondary literature.

The science of conservation biology is developing fast, and its scope continues to diversify. In essence, *any* biological knowledge of invertebrates can be relevant to some aspect of their conservation, and one problem with this book has been to decide how much 'basic biology' to include, especially on their life cycles, ecology, and the factors limiting their natural distributions. Overviews of these and other topics are readily available from standard texts, so I have confined information given here to points relevant to interpreting and understanding particular examples discussed in the text. Sound conservation management invokes sound biological understanding. In the allied science of insect pest control, the need to understand the pest species' life system and population dynamics has long been recognized as essential to optimal pest management: the need is equal for management of a species for preservation (rather than destruction), but the opportunity to obtain equivalent insight and background knowledge is less, because of the lack of tangible economic benefit or returns for the investment involved. The urgent requirements are to:

(1) increase the funding and scientific capability for the basic studies necessary to conserve invertebrates and assess their roles as indicators of environmental health; and

(2) apply that support in the most effective ways possible, in practical conservation programmes to safeguard the greatest possible array of taxa and the habitats that sustain them.

Melbourne
June 1994

T. R. N

ACKNOWLEDGEMENTS

In preparing this book I have been very aware of the tolerance of colleagues on whom I have inflicted ideas or opinion, or who have helped me to gain some perspective on invertebrate groups and communities which I have not studied directly. In particular I would like to acknowledge the stimulation of many members of the World Conservation Union Species Survival Commission's Invertebrate Conservation Task Force. Dr A. L. Yen critically read an advanced draft, and I appreciate greatly his constructive appraisal, and an earlier version benefited from the sensitive editorial advice of Oxford University Press and the detailed comments of two referees. Mrs Tracey Carpenter, Mrs Rhonda McLauchlan, and Ms Jenny Browning have patiently typed successive drafts, and Mrs Carpenter also prepared most of the figures.

The following publishers and authors are thanked for permission to reproduce or modify diagrams or tables: American Fisheries Society; American Institute of Biological Sciences; Blackwell Scientific Publications Ltd; Cambridge University Press; Chapman & Hall Ltd; Council of Europe; CSIRO Editorial Services; Ecological Society of America; Elsevier Science Publishers; Foundation for Environmental Conservation (Prof. Dr N. Polunin); Gustav Fischer Verlag; Kluwer Academic Publishers; Landcare Research N.Z. Ltd (Dr M. J. Meads); C. V. Mosby Publishers; Pergamon Press Ltd; Prof. M. J. Samways; Society for Conservation Biology; Surrey Beatty & Sons Ltd; University of Hawaii Press, Honolulu.

CONTENTS

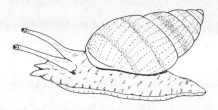

1 INTRODUCTION: BIODIVERSITY AND INVERTEBRATES

This chapter introduces the major themes discussed in the book, noting the abundance and variety of invertebrate animals and the difficulties of appraising this diversity. Difficulties occur even in establishing broad patterns of species numbers over most of the world, or assessing the relative diversity of terrestrial and marine communities. This massive uncertainty is difficult to overcome, yet doing so is a central—probably the most important—theme in increasing our knowledge of the Earth's biota and leading to effective planning for their conservation.

BIODIVERSITY

The term 'biodiversity' encompasses all levels of biological variation, from the level of genes to ecosystems, but is most commonly interpreted at the level of organisms—simply, the numbers of different kinds of animals or plants: species. This is usually the easiest way to measure diversity. The most numerous kinds of animals are largely ignored in conservation planning, and many people who are very sympathetic to conservation of 'biodiversity' tend to regard these, the invertebrates, as of little relevance or importance to the sustainable well-being of the natural world. This is far from the truth. Animals such as corals, earthworms, snails, and insects may not engender public sympathy or interest to the same extent that, for example, large mammals or rare birds do, but they may have much more important roles in the integration and working of natural communities

and sustaining these when they are disturbed. There are exceptions to this widespread attitude though. Butterflies, in contrast to most other insects, are one group of invertebrates which people generally 'like', and which many equate with vertebrates in importance and aesthetic appeal. Indeed, ignorance of invertebrates (rather than simple prejudice) may be at the root of much 'vertebrate chauvinism' and increasing peoples' realization of invertebrate importance is a critical factor in overcoming this. This book is about these other animals, and is an attempt to illustrate the importance of invertebrates, the needs and principles for conserving them in the face of accelerating, continuing, and ever-diversifying threats to their existence, and the emerging concern for their future.

Many biologists predict that a substantial proportion of the world's species will be exterminated within a few decades in an unprecedented mass extinction which is eloquent and tragic testimony to the needs and desires of burgeoning human populations. May (1988) stated that 'conservation biology is a science with a time limit', and stressed that current and predicted rates for species extinctions exceed by far the usual rates of species generation, whereas these two processes have been approximately equivalent over much of the past.

The arguments for conserving biodiversity (McNeely *et al.* 1989) include:

(1) that each species is part of a holistic ecosystem and the loss of any part (species, population, etc.) may lead to instability and collapse of the whole system;

(2) that conserving species may help to maintain broader 'life support systems';

(3) that species are, or may in the future be, sources of products useful to people, or in facilitating human well-being; and

(4) that diversity has scientific and cultural values, as well as aesthetic appeal.

Any of these considerations applies significantly to invertebrates, and they are encapsulated in Wilson's (1987) statement that 'if invertebrates become extinct, the world as we know it would cease to exist'. What then, *do* we know about invertebrate diversity?

INVERTEBRATE DIVERSITY

The vertebrates (fish, amphibians, reptiles, birds, and mammals) comprise only one major group (phylum) of the animal kingdom. By contrast, the simple term, 'invertebrates' unites a great diversity of life forms and unrelated animal groups by the single condition of lacking a backbone, and represents a highly artificial amalgamation of evolutionary lineages. Many of the 25–30 (or so) major invertebrate groups, or phyla (Table 1.1), are taxonomically isolated, and their grouping under a convenient 'umbrella term' must not be taken to mean that only two major categories exist in the animal kingdom. The invertebrate animals are immensely varied and occupy, collectively, the totality of global ecosystems and represent virtually all life styles and trophic roles. In this book, I am excluding discussion of the single-celled organisms, the Protozoa, because of the emerging general consensus that these constitute an evolutionary level and structural plan that is fundamentally different from the multicellular 'true animals'. Protozoa show immense adaptive radiation, and some recent evaluations suggest that they comprise many different phyla of unicellular beings.

The twin roles of invertebrates in conservation studies are:

(1) as targets for conservation *per se*; and

(2) as tools used to monitor the 'health' of natural environments, to indicate the effects of various anthropogenic and other intrusions, and to show the success (or otherwise) of remedial management procedures.

Knowledge of the various multicellular (metazoan) invertebrate groups is very uneven, and it is not currently possible to appraise the conservation needs, or the roles in conservation management, of many of these. This book reflects this patchiness and, inevitably, draws most of its content from the better-known and more completely documented taxonomic groups. Members of very few of the phyla (Table 1.1) have played either of the above roles.

The future implications of this are discussed in the final chapter. I have tried to include examples and discussion bearing on many phyla, but it is important to confess that devotees of many invertebrate phyla will find little here that deals specifically with the vulnerability or conservation of their favourite groups, simply because that information or concern does not exist or has not been documented. The challenge remains to fill these gaps and to determine their conservation needs, other than the universal need to preserve suitable habitats from the increasing array of external threats. These lacunae extend to some of the largest groups of animals. Nematode worms, for example, have scarcely been used in conservation, despite their ubiquity. Some recent work is demonstrating their values as monitors of toxins in the sea, and as possible measures of soil community maturity.

Indeed, one could validly claim that habitat conservation is of such overriding importance to invertebrates that concentrating on taxa to any extent is trivial—tinkering around the edges of the problem. Yet the most tangible way to understand the habitat needs for animals or plants is through a study of the organisms themselves, and the question then becomes one of whether particular taxa may yield more, and more valuable, general information than others.

Another major bias of this book is, perhaps, equally inevitable. The variety of major invertebrate life forms in marine environments is

Table 1.1 The invertebrates: major groupings, habitats, and diversity (after Meglitsch and Schram 1991)

Phylum	Common names	Major habitat(s)	Number of species known/estimated
Mesozoa	(Mesozoans)	Microscopic parasites	85/500+
*Porifera	Sponges	Aquatic, mainly marine	$c.$ 5000/?
*Cnidaria	Corals, polyps, sea anemones	Aquatic, mainly marine	$c.$ 10 000/?
Ctenophora	Comb jellies	Marine	80/130–500
Gnathostomulida	(Gnathostomulids)	Microscopic; Interstitial in sand, marine	80/1000
*Platyhelminthes	Flatworms, flukes, tapeworms	Aquatic or terrestrial; many parasitic forms	/many 1000s
Gastrotricha	(Gastrotrichs)	Aquatic, interstitial	450/1000+
Rotifera	Rotifers, wheel animalcules	Aquatic	1800/ 2500–3000
Acanthocephala	Thorny-headed worms	Intestinal parasites	$c.$ 1000/?
Loricifera	(Loriciferans)	Marine sediments	?/?
Kinorhyncha	(Kinorhynchs)	Aquatic, interstitial	125/500
Priapulida		Marine, bottom dwellers	15/25
Nematomorpha	Horsehair worms	Aquatic, parasites	275/?
Nematoda	Roundworms	[All]	1×10^6/?
Chaetognatha	Arrow worms	Marine	70/100
*Mollusca	Snails, clams, octopus (etc.)	[All]	100–150 000/?
Nemertinea	Ribbon worms	Mainly marine	800/3000
Sipuncula	Peanut worms	Marine	320/330
Echiura	Spoon worms	Marine	130/140
Pogonophora	Beard worms	Marine	120/500
*Annelida	Earthworms, leeches polychaetes	[All]	12 000/?
Pentastomida	(Pentastomids)	Parasites of vertebrates	$c.$ 100/?
Tardigrada	Water bears	Mainly semiaquatic	550/1000+
*Onychophora	*Peripatus*	Terrestrial	100/300
Arthropoda	Insects, crustaceans, arachnids, and their relatives	[All]	1×10^6/ $?10–30 \times 10^6$ (or more)
Phoronida	[Phoronids]	Marine	13/20
Bryozoa	Entoprocts, ectoprocts	Mainly marine	4000/?
Brachiopoda	Lamp shells	Marine	330/500
*Echinodermata	Starfish, sea urchins	Marine	8000/?
Hemichordata	Acorn worms	Marine	100/150+

Numbers for diversity are usually very approximate.

* Phyla to which conservation effort has been directed, and which include most of the examples discussed in this book.

probably even greater than that in terrestrial and freshwater environments, and a number of significant phyla are entirely marine. Yet, other than in estuaries and littoral environments, shores, and coral reefs, little attention has been paid to their conservation needs in relation to the greater notice accorded invertebrates of terrestrial and freshwater ecosystems. Other than for commercially desirable taxa, such as euphausid crustaceans (such as krill, p. 23), in general the inaccessibility and lack of knowledge of the deep-sea oceanic assemblages has restricted such appraisal, other than general warning of the potential hazards of dumping industrial and other wastes into the sea. Lack of 'equal time' for marine invertebrates in this book does *not* reflect unequal importance but simply the current impossibility of appraising the status and conservation need of most taxa there, other than by bland platitudes. Likewise, lack of examples of many of the other invertebrate taxa should not be taken to imply that they are unimportant or 'expendable' but, as we shall see, there are severe practical barriers to evaluating the conservation biology of many of these forms.

The diversity of invertebrate life forms is summarized in Table 1.1. Structural and biological features of these are appraised in many excellent textbooks (such as Marshall and Williams 1982; Meglitsch and Schram 1991), but the detailed ecology of many of the phyla is known only rather sketchily. The table emphasizes that relatively few groups, indicated by an asterisk, have received direct conservation attention, and for most of these only a few species have been appraised in this way.

The diversity and biomass of some of the invertebrate groups defies accurate appraisal. Consider, for a moment, Buchsbaum's (1966) introductory quotation to his chapter on nematode worms:

If all the matter in the universe except the nematodes were swept away, our world would still be dimly recognisable, and if, as disembodied spirits, we could then investigate it, we should find its mountains, hills, vales, rivers, lakes and oceans represented by a film of nematodes. The locations of towns would be decipherable, since for every massing of human beings there would be a corresponding massing of certain nematodes. Trees would still stand in ghostly rows representing our streets and highways. The location of the various plants and animals would still be decipherable, and, had we sufficient knowledge, in many cases even their species could be determined by an examination of their erstwhile nematode parasites.

The evocative scenario is one of nematode ubiquity, an enormous variety of specialized roundworms associated with all other life on Earth, many very specific in habit, but collectively a daunting diversity of species and vast biomass. Yet, arguably, nematodes are not even the most speciose of invertebrate groups (a status usually accorded to the Arthropoda) and we cannot merely dismiss the above quotation as hyperbole, because we know too little about nematodes. Some authorities claim that only a few per cent of all nematode species have yet even been named by scientists, and similar comments could be made for some other large phyla. Furthermore, many species can live in small areas of the world, and form complex associations in very restricted habitats. Thus, about 340 species of flies are adapted to living in the beech forests of central Europe, and a single old stump may harbour larvae of up to 20 different families (Walter and Breckle 1989); a small German stream contained more than 1000 invertebrate species (Allan and Flecker 1993), and more than 800 benthic species occur in 10 m^2 of Bass Strait, separating Tasmania from mainland Australia (Poore and Wilson 1993).

Some of the figures for species diversity in Table 1.1 are clearly approximate and uncertain. Indeed, some differ substantially from other, equally informed, recent estimates. For example, Groombridge (1992) cited the following numbers of described/estimated total numbers for some groups; nematodes (15 000/0.5–1.0 million), molluscs (70 000/200 000), crustaceans (40 000/150 000), spiders and mites (75 000/0.75–1.0 million), insects (950 000/8–100 million).

Invertebrate importance extends well beyond simple numbers of species. In forest soils of South America, the collective biomass of the few hundred species of social insects (predominantly termites and ants) exceeds by far that of vertebrate animals in the forests. New Zealand soils, to 10 cm deep, can support several million nematodes a square metre (Yeates 1994). In some New Zealand forests, a single introduced species of social wasp (p. 63) is claimed to eat more insects than the entire resident population of insectivorous birds. Insects, a major component of the phylum Arthropoda, are generally presumed to be the most speciose class of animals, and several attempts have been made to estimate the number of existing species. There is still considerable uncertainty and, from some points of view, the question is of little more than academic concern. However, trying to estimate and understand the organismal level of biodiversity, its global and ecological patterns, and how it is influenced by human activity and other disturbance is the core of conservation, in providing the information necessary to 'manage' and restore the natural world. These patterns, essentially, are the template for measuring conservation need. It is thus important at least to appreciate the limitations and uncertainties with which we are faced in estimating numbers of extant species.

The two main approaches to obtaining these data are:

(1) extrapolation from relatively small or local samples of animals which have been analysed in detail and which are assumed to be representative of the community or area sampled; and

(2) seeking a consensus of opinion from the most knowledgeable experts in the groups involved.

These have sometimes led to vastly different estimates for some invertebrate groups, and the grounds for extrapolation from samples or from localized knowledge are sometimes flimsy. Grassle (1991) and Grassle and Maciolek (1992) extracted marine animals from bottom sediment samples covering 21 m² of the north-eastern Atlantic continental shelf. They found 798 species of invertebrates and concluded that extrapolation to estimate species diversity in the deep seas could be made. Their conservative derived estimate of some 10 million species was likened

by Briggs (1994) to determining the terrestrial species diversity for North America by counting organisms in samples of soil from a backyard. But diversity in Grassle and Maciolek's samples was at the lower end for a suite of samples of benthic fauna from different oceans and latitudes (Poore and Wilson 1993). The expert opinion approach was exemplified for insects by Taylor (1976) and Gaston (1991) but poor information can lead to unsubstantiated estimates, and there seems to be no real substitute for practical information based on adequate series of samples.

An important aspect of conservation is communicating concern to managers who may be able to influence patterns of resource exploitation and disturbance to natural environments, and parameters such as 'high diversity' or the presence of 'notable' species are persuasive features for promoting conservation measures directed at the environment or the species itself (p. 73). The converse situation, *not* being able to quantify diversity of invertebrates, is taken frequently to reflect lack of interest or importance. Many decision-makers find it difficult to understand and accept the contrast between the complete (or almost complete) inventories or species lists which are usually achievable easily for birds, mammals, and vascular plants, in particular, for a given site, and uncertainties over, even, orders of magnitude for many invertebrates in the same localities.

Global distribution of invertebrate diversity is not even. Despite the comment on numbers of phyla on p. 2, it is still debatable whether terrestrial invertebrates are more or less speciose than marine ones, for example. The most species-rich terrestrial environments are generally assumed to be tropical forests, but Platnick (1991) believed that a truer global pattern is of numerical predominance in the tropics and the southern temperate regions, with Arthropoda being by far the predominant component, so that the global picture of terrestrial species numbers would be 'pear-shaped', with the narrow part in the northern hemisphere.

As well as changes in diversity estimates for relatively well-known invertebrate groups, Barnes (1989) noted that several major taxonomic categories of invertebrates have been discovered only in the past 30 years or so (the marine phylum Loricifera, a class of echinoderms, four classes of crustaceans, extant Monoplacophora (Mollusca), and an anomalous 'worm': *Lobatocerebrum*), and others have become familiar only over this period (examples: the beard worms, Pogonophora; the microscopic Placozoa; pterobranch hemichordates). Biological knowledge of all these is fragmentary and unlikely to be improved substantially in the foreseeable future. Some are minute organisms living in habitats such as the ocean depths, which are generally inaccessible to biological study. Such discoveries counsel against complacency that even all major evolutionary lineages and groups of invertebrates are known, let alone the uncertainties at lower taxonomic levels.

ASSESSING DIVERSITY

Some recent estimates for numbers of taxa in the main invertebrate phyla are listed in Table 1.1. There is substantial uncertainty in some of these estimates, as noted above. The major hindrances to obtaining more accurate knowledge are:

1. Poorly documented faunas, especially for small and inconspicuous animals. Even in the countries in which there is a strong tradition and history of faunal documentation, such as parts of Europe and North America, there are no definitive species-level synopses for many invertebrate groups. For much of the tropics, in particular, the lacuna yawns large: even reasonably comprehensive *collections* of many invertebrates have yet to be made systematically, let alone studied in detail for faunal documentation. It is common for a specialist examining a collection from the tropics (including nearly all the 'megadiversity' areas of the world) to find a very high proportion of previously unknown taxa. Soberingly, it is not unusual to find that the habitats from which these animals have been collected have been destroyed, so that many of the species may already be extinct when they are recognized. This destruction is widespread: a Californian grasshopper was recently given the

specific epithet of '*extincta*' to draw attention to the loss of its dune habitat.

2. There are very few specialists or devotees working on some of the groups, including some major taxa which are known to respond subtly to environmental changes and potentially useful as indicators of environmental quality (p. 31). There is thus little realistic chance of documenting the fauna comprehensively, especially in tropical areas, because most systematists are based in the temperate regions of the world.

3. Most taxa are examined and described from dead museum specimens (not always in optimal condition), often from very few specimens. Morphological differences between taxa may be very small and masked by variability, and examination of more comprehensive fresh material may lead to different conclusions on the limits of species. Chemical and biological (including behavioural) differences may occur between populations of morphologically similar organisms, and the limits of species are commonly very difficult to detect, as exemplified below. Likewise, the variability within species is a dimension of diversity, and loss of isolated populations, even of widespread species, is a proper conservation concern in reducing evolutionary potential.

4. Many invertebrates are short-lived, or are quiescent (for example, as eggs) for much of the time. The distribution of most taxa, even conspicuous ones, is difficult to define except by long-term studies and surveys. Unlike a tree or a resident bird, which is conspicuous for the whole year, many insects are present as adults for only 2 or 3 weeks (or, even, less) each year in temperate regions.

Three groups of invertebrates suffice to exemplify many of the difficulties involved in assessing what might be regarded as one of the most fundamental aspects of knowledge, the numbers of species present. Similar comments could be made on nearly all invertebrate groups.

Water mites

Abundant freshwater organisms, Hydracarina have recently received considerable specialist attention (for example by Cook 1974, 1986), with the result that hundreds of previously unknown species are being recognized and described from many parts of the world, different sexes associated for the first time, and juvenile stages associated with the corresponding adults.

Our general level of knowledge of the systematics and diversity of water mites is thus being raised substantially simply because of increased attention from a few systematists, but recognition of juvenile stages is of much wider relevance. In many invertebrate groups, the immature stages undergo a dramatic change in form, metamorphosis, to attain the adult stage, and larvae and adults differ greatly in appearance and biology. A maggot and a fly, a zoea and a crab, a cercaria and a fluke are each very different, and adults and juveniles of most species have not been associated unambiguously by rearing in captivity. Interpreting the number of species in samples from wild populations of such animals is fraught with difficulty, as juvenile and adult stages may both be diverse.

Velvetworms: Onychophora

This group, *Peripatus* and its allies, comprises 'living fossils' of debatable evolutionary relationships, with some recent work (Ballard *et al.* 1992) placing them within the broad arthropod assemblage. Onychophora has conventionally been regarded as one of the least diverse and best-documented invertebrate phyla, with rather fewer than 100 species described and most taxa having very circumscribed distributions. However, recent studies on the genetics and chemical features of the Australian and Caribbean Onychophora have revealed unexpectedly high levels of variation and numbers of species. In Jamaica, for example, two genetically distinct forms masquerade under the name *Plicatoperipatus jamaicensis*, sometimes in different sites but elsewhere with populations which coexist but do not interbreed (Hebert *et al.* 1991). For the Australian fauna, it seems that well over half the onychophoran species revealed by chemical features have not yet been described in conventional morphological terms (Tait *et al.* 1990), with the attendant interpretative problem (visible only with hindsight!)

that much historical documentation of particular species may be unreliable because of confusion between distinct forms which were not then recognized.

The incidence of such 'cryptic species', which are scarcely—if at all—separable on morphological characters but are distinct biological entities, occurs also in other invertebrate groups, and may be very common. It has been estimated as possibly as high as 2.5 times the number of 'obvious' species in the Canadian insect fauna, for example (Danks 1979).

Insects in tropical forests

Erwin's (1982) study of neotropical forest beetles, more than any other single study, has engendered and stimulated debate on the magnitude of invertebrate diversity. He was able to use insecticide-fogging techniques, whereby the previously inaccessible rich fauna of the forest canopy could be sampled by misting it with non-persistent pyrethrin insecticides from machines hoisted into the canopy on ropes, and catching the rain of falling invertebrate bodies on sheets or funnels suspended near the ground. This technique revealed massive numbers of previously unknown species. Erwin captured more than 1100 species of beetles from 19 trees of a single species, *Luehea seemannii*. Estimating that 20 per cent of the herbivorous species were specific in some way to the tree species (160 beetle species), that beetles constitute 40 per cent of known arthropod species, that the canopy fauna is at least twice as rich as the forest-floor fauna and largely different from it, and using an approximation of 50 000 species of tropical trees, Erwin arrived at a suggestion of possibly 30 million tropical arthropod species. Each of his approximations is open to question, but the study has been of immense importance in indicating an approach to getting 'real figures' for organismal biodiversity.

Stork (1988, 1991) used an independent set of canopy invertebrate samples from Borneo forests to discuss and compare Erwin's interpretation. Depending on the validity and level of the assumptions made, estimates of arthropod species numbers could go as high as 80 million! Many authorities believe that a more realistic estimate is in the range of 5–10 million but, as May (1988) emphasized 'we do not know to within an order of magnitude how many species of plants and animals we share the globe with'.

INVERTEBRATES IN CONSERVATION

This massive uncertainty continues to foster debate (for example, Erwin 1991; Gaston 1991), sometimes referred to as 'the biodiversity debate'. The maxim of needing to know 'what is there' in order to be able to understand and conserve any local biota in a habitat pervades much conservation planning, and the impracticality of being able to produce complete or near-complete invertebrate inventories or, even, defining local diversity within close limits, must be appreciated. In his thought-provoking essay on biodiversity conservation, Vane-Wright (1994) reminds us of Aldo Leopold's maxim, that 'the first rule of intelligent tinkering is to keep all the bits'. In interfering with natural systems, we need to be able to define the 'bits' (species). Complete 'species lists' for most invertebrates are currently impossible to produce, even for small areas or habitats. Approaches to this in poorly documented faunas in the tropics have fallen largely into one of two categories:

(1) to 'get them all' by complete/near-complete or intensive collecting; or

(2) to 'sample and estimate', extrapolating from suites of samples to estimate what the total number of different taxa may be, assuming that the samples are fully representative.

Close approximations, even definitive listings, can be obtained for a few well-known groups, such as butterflies, dragonflies, and terrestrial molluscs, in temperate regions and in a (very) few intensively studied tropical sites, but little or no data are likely to exist for many of the abundant invertebrate groups even in relatively well-documented temperate regions. Prospects for ameliorating this depressing situation are remote. However, the recent global initiative known as 'Systematics Agenda 2000' (SA 1994)

has the major objective 'To discover, describe and inventory global species diversity' from the basis of a massively increased research effort over the next 25 years. How this research effort is to be increased adequately is as yet uncertain.

Unlike most invertebrate assemblages in other regions, it may indeed be possible to make definitive inventories of the relatively low-diversity Antarctic assemblages at a number of different sites, to assess the value of these for conservation priority (Usher and Edwards 1986). Criteria for site evaluation in the Antarctic are more restricted than for most other parts of the world—involving representativeness, uniqueness, type localities, public appeal (areas involving breeding colonies of birds or mammals) and typicalness. Application of such criteria to low-diversity arthropod assemblages, such as those sampled by Usher and Edwards, where no more than 17 species (of an overall total of 25) were found at any one of the seven sites examined, is difficult because of local variability. 'Uniqueness', by which maximum protection is afforded to sites whose biota are not known to occur elsewhere, could be justified on the arthropod records obtained. 'Representativeness' was the most important criterion in the Antarctic, so as to conserve examples of all kinds of assemblages, whereas 'uniqueness' ensures the conservation of important exceptions to this. This principle is of much wider application in designation of priority sites for reserves in other parts of the world, and the criteria noted (after Ratcliffe 1977) have been used widely in Britain for evaluating sites on the basis of the taxa present.

However, the general level of diversity and the incidence of 'notable' invertebrate taxa are both invaluable conservation indices (p. 31). Particular kinds of invertebrates can be highlighted as priorities in particular contexts (p. 22) and studies on them can yield detailed information useful also in appraising ecologically similar taxa.

Nevertheless, lack of knowledge of the groups or taxa present in an area or habitat and, therefore, of the influences of various external factors on the amount and pattern of diversity and numbers, can sometimes be a serious impediment to promoting and implementing conservation. Lack of knowledge is equated, all too frequently, with lack of interest or importance. The major knowledge gaps relate to taxonomy and ecology, and are commonly termed 'impediments' to facilitating invertebrate conservation. Both are enhanced by the small size of many invertebrates. Not all are minute (the marine giant squid, *Architeuthis*, can attain a length of 18 m), but many are commonly overlooked, and specialized equipment or treatment is needed to collect and define them.

The taxonomic impediment to understanding invertebrate diversity

That many invertebrates do not have scientific names yet (what Taylor (1976) has termed the 'taxonomic impediment') leads to difficulty in effective communication, but this situation is unlikely to be overcome without a massive renaissance in descriptive taxonomy and increase in the taxonomic workforce: some aspects of this need are discussed in Chapter 8. It is paradoxical indeed that at a time when more and more people are stressing the urgency and need to document natural assemblages, almost all of which are dominated by invertebrates, as a basis for informed conservation and sustainability of natural ecosystems, experts in the world's major taxonomic institutions are being retrenched or forced to change the orientation of their work and are not being replaced, and such very basic science is coming to be regarded increasingly as unfundable, despite strong advocacy from many concerned biologists. Even if the formal naming of invertebrates is omitted (p. 149), it is commonly possible only for a specialist in a particular taxonomic group to appraise a collection properly, to determine the approximate numbers of taxa present, and to state which of these are 'exciting', significant, or of particular value in conservation assessment.

Such expertise reflects many years of accumulated wisdom and experience. It, emphatically, is not possible for an inexperienced person to fulfil this role, or even to recognize and diagnose most species confidently. With appropriate training, though, 'biological diversity technicians' (sometimes termed 'parataxonomists') can become proficient in the sorting, preparation, and broad-scale recognition of taxonomic groups for comparison of samples. Their role is important in helping to overcome the related 'ecological impediment' (below) but cannot substitute effectively for higher-level expertise. A national biological survey organization (the Instituto Nacional de Biodiversidad: INBio) in Costa Rica, for example, utilizes parataxonomists to collect and prepare enormous numbers of arthropods from field surveys (Alberch 1993), to represent and inventory the country's invertebrate diversity, but many groups included in those accumulations await specialist attention.

Distribution of taxonomic expertise is highly uneven, both among taxa (many large groups of invertebrates have fewer than a handful of specialists who understand anything of their systematics) and geographically (so that greatest concentrations of taxonomic expertise are found in temperate regions of the world, where invertebrate diversity may be low in relation to much of the tropics).

Ecological impediments to understanding invertebrate diversity

Systematic knowledge is extremely important also in ecological interpretation and determining the functional state of natural communities. In any community, the invertebrates present comprise a gradation of ecological forms and roles. As well as each having a particular trophic role (for example as a herbivore, predator, or decomposer) and set of interactions with other species present, their influences in the community will differ. Some will be relative 'generalist' species, usually with broad habitat and resource tolerances (loosely, with broad niches), which may be relatively insensitive to moderate disturbance, as they can utilize alternative resources if one kind is affected adversely. Many others will be relative 'specialists', with narrow, specialized resource needs, which depend very precisely on those resources (such as a particular species of foodplant or animal food) being available. In general, a high proportion of specialist species is characteristic of more mature or late successional habitats, and a higher proportion of generalist species occurs in early successions or more disturbed habitats or sites. This ecological spectrum or 'habitat templet' (Southwood 1977, Greenslade 1983) is of considerable relevance to conservation, and reflects that the quality and characteristics of a habitat influence the kind of species that can live there. Highly disturbed habitats may still support a high diversity of species, but may have lost those ecological specialists of great evolutionary significance.

Thus, many invertebrate species persist in, or newly colonize, highly disturbed areas so that diversity there can appear high and, without further appraisal, the system might appear not to need conservation. However, more detailed appraisal might well reveal that many of the ecologically sensitive specialist species characteristic of a pristine habitat of the kind originally present had indeed disappeared and had been replaced by numerically abundant 'weedy' species, those fecund generalists which can thrive under a wide range of conditions.

In general terms, a high diversity of species (or species richness) has been equated to a 'healthy' environment and any marked diminution in species richness as a cause for concern. Species diversity is one of several indices valuable in comparing and ranking communities or sites for conservation value. Clearly, the fundamental capability to assess 'diversity' in terms of numbers of species present is of central importance to such rankings. Conservation concern relates also to the susceptibility and vulnerability of specialist taxa to environmental changes. Most of the invertebrates that have been the subjects of specific conservation concern are such specialized species.

The ecological impediment to invertebrate conservation reflects diversity, the difficulties of sampling it, and of interpreting the content and meaning of the sampled material adequately. This theme will be considered more fully in Chapter 4, but pervades much of the rationale of invertebrate conservation in relation to logistic constraints, and so needs introduction here. Virtually any sample of 'biodiversity' will contain a massive number of invertebrate animals which will need to be preserved, sorted, and prepared carefully before they can be studied or appraised properly. Many small animals need

to be mounted on microscope slides, for example, to identify them even to higher categories reliably. The costs of this necessary routine are high and the needs often not appreciated adequately by funding bodies. Cost estimates of preparing and identifying invertebrate specimens vary considerably but, at the higher end of the range, can exceed US $100/specimen. Yet such preparation and appraisal are only the initial steps in assembling the more relevant data, the synthesis of the interdependence between the species found, their integration into communities or looser assemblages, knowledge of the resources on which each may depend, interpretation of life cycles, and so on. *Any* ecological knowledge is likely to have conservation relevance, but even basic details of biology are unlikely to be available for most, perhaps almost all, species present in a sample. Interpreting such samples in ecological terms is, therefore, difficult. Quantitative appraisal of invertebrate communities or local faunas is by no means a simple task: gaining understanding of biological systems in order to conserve or manage them is difficult and expensive.

Much recent emphasis in using invertebrates in conservation has been on seeking useful or productive short cuts through this logistic morass, particularly via seeking particularly informative taxonomic groups or ecological interactions, and less labour-intensive sampling methods for invertebrates than have been employed in many earlier studies. These constraints apply particularly to free-living invertebrates. The incidence of many specialized forms which are internal or external parasites can be studied only by collecting the appropriate hosts. Their well-being depends entirely on that of the host animal or plant, an obligatory resource for a specialized parasite or other monophagous consumer.

There is little specific information on the decline of parasitic groups of invertebrates, though any decline in hosts will, of course, affect them. Most available information is on ectoparasites of vertebrates and, as Rózsa (1992) has emphasized, far more attention is likely to be paid to the conservation needs of the host than of the parasites themselves. Thus, the pygmy hog sucking louse, *Haematopinus oliveri*, occurs only on the endangered pygmy hog in northern India, which is threatened by destruction of its savannah habitats (Wells *et al.* 1983). The fluke *Proterometra dickermani* (Azygiidae) infests the snail host *Goniobasis livescens*, and is known to occur only in the Ocqueoc River, Michigan. The percentage of infested snails decreased progressively from 85 per cent in 1980 to only 4 per cent in 1989, and snails feed naturally on the algal coating on submerged rocks and gravel (Uglem *et al.* 1990), a food supply which is eliminated by siltation. Siltation, resulting from erosion from adjacent croplands, also increases egg mortality and hampers snail-to-snail transmission of the fluke. Sea lamprey control in the river by use of pesticides might also have contributed to the decline of snails, because the lampricide used was often applied with a molluscide to improve its effect. *Proterometra* is regarded as an endangered (p. 169) species (Uglem *et al.* 1990).

The correlation between large host population size (abundance, range area) and numbers of any ectoparasite species might be a general one. Large host populations tend to harbour more parasite species than do smaller ones, and with decline of the hosts many parasites might become extinct (Rózsa 1992).

BIODIVERSITY AND CONSERVATION BIOLOGY

The different levels of 'biodiversity' reflect different levels at which 'biological conservation' may be defined, with the overall aim of maintaining the hierarchy from genes to biosphere (Fig. 1.1), including the diversity of organisms and their habitats with the intricate interrelationships which occur amongst and between them. Conservation effort, thus, may be directed at populations or demes, single species, larger groups of allied biota, or the whole mass of living organisms. Conservation biology, simply, is the biological information needed to fulfil such aims and its organization and implementation as practical management programmes. It incorporates taxonomy, genetics, ecology, distributional status, and population dynamics into assessing the needs for conserving a taxon or system and

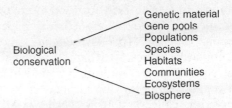

Fig. 1.1 The various levels at which biological conservation can be directed, ranging from particular genetic variants, perhaps manifest in only small proportions of a population, to the global biosphere.

to pursuing actions or management needed to maintain or restore that entity.

It is not feasible to include all kinds of invertebrate animals specifically in an appraisal of the conservation biology and needs of such a vast and heterogeneous group. Simply, our knowledge of the biology and distribution of some taxa is so poor, or their systematic framework is so impoverished and devoid of specialist attention, that constructive synthesis is effectively precluded (Chapter 8). However, the general themes and principles emerging from study of the better-known invertebrate groups are generally applicable, but, inevitably, some constraints to a total picture are imposed by ignorance of an ecological diversity extending from the abyssal oceans to the high mountains and in which every nuance of ecological role is included.

Conservation need manifests most commonly as a response to a perceived decline of a habitat or of the abundance and distribution of one or more species. For invertebrates, the latter has most commonly been to increasing scarcity of highly visible taxa (such as butterflies) and the development of insect conservation, in particular, includes many such cases (Samways 1994*b*). Such declines are associated frequently with anthropogenic changes, predominantly direct habitat destruction or pollution, and may be very difficult to distinguish from natural decline or long-term population fluctuations. A given animal population is continually variable in the number of individuals it contains. Births and deaths occur throughout its history, and many invertebrate species are characterized by high-amplitude fluctuations over periods of many generations, sometimes associated with weather

(Lepidoptera: Beirne 1955) or food availability. Many species, however, are characteristically 'common' or 'rare' (p. 76), but it may be extraordinarily difficult to determine whether perceived rarity is natural, or induced by human activity. The latter is the proper cause for conservation concern at the species level. In nature, many species become extinct if they lack the capability to adapt to changing conditions over time, perhaps by evolving to form new species. Many anthropogenic changes to natural ecosystems, especially during the past century or so, have been so rapid and severe that they have effectively removed much of the option for such gradual adaptation to changed conditions, so that extinction has become the major or sole option for many ecologically sensitive plants and animals. Maintaining habitats or biomes is the key to preventing or slowing extinction.

Declines may be in numbers, in geographical range, or both, so that increasing rarity and narrowing distribution are both involved.

Despite the certainty that human activities are currently causing massive extinctions of species, documented cases of invertebrate species extinctions are rather rare, and their causes often intricate and difficult to clarify. It is not always easy to attribute decline or extinction of an invertebrate to any specific cause, even when that decline has been spectacular. The Rocky Mountain grasshopper (*Melanoplus spretus*) was the most serious agricultural pest in western North America before 1900, and vast migratory swarms destroyed crops over much of that large area during the nineteenth century. *Melanoplus* declined rapidly in the late 1800s and the last living specimen was collected in 1902. Its disappearance has been described as 'one of the most compelling unsolved ecological and entomological mysteries of our time' (Lockwood and De Brey 1990). Indeed, this instance may represent the only extinction of a severe pest species in the history of agriculture. Lockwood and De Brey attribute this to agricultural activities, leading to destruction of favoured rangeland habitats and increased numbers of insectivorous birds. The spread of railroads accelerated human settlement, and development of agriculture in the western United States, and the period of

precipitous decline of *Melanoplus* coincided with this rapid habitat alienation and introduction of exotic species, widely recognized as among the major threats to invertebrates in many parts of the world (Chapter 3).

The eelgrass limpet, *Lottia alveus*, is one of rather few documented marine invertebrate extinctions, and was once distributed widely along the Atlantic coast of North America. The last known living specimens were collected in 1929, and there is little doubt that it is now extinct (Carlton *et al.* 1991). The decline coincided with the massive loss of eelgrass (*Zostera*) from 1930 to 1933 and, although some eelgrass survived in lower-salinity waters, it seems that the limpet may have had a very narrow range of salinity tolerances so that it could not survive in those possible refuge areas.

Many (most) such declines and extinctions in invertebrates pass unheralded and unnoticed. The exceptions tend to be declines in species which are in some way 'prominent'—such as those used as human foods (molluscs, some crustaceans), in medical treatment (medicinal leeches), for commercial sale (corals, sponges, molluscs, butterflies), or with popular aesthetic appeal (butterflies), collectively a very small proportion of the total. The tradition of concern for invertebrate species is well-founded on butterflies, especially in Europe (Kudrna 1986, New 1991*a*). One of the earliest classic cases is of the British large copper butterfly, *Lycaena dispar dispar*, which became extinct around the middle of the nineteenth century due to a combination of habitat destruction (drainage of fenland) and intensive exploitation for commercial sale to collectors (Duffey 1977). The introduction to Woodwalton Fen of a continental subspecies (*L. d. batava*) was made initially in the late 1920s, and regular releases of captive-bred stock have been made since that time. Without such augmentation it is very doubtful whether the large copper could survive in the wild in Britain. Declines of other butterfly species and, more rarely (simply because they have not been documented to the same extent), of representatives of other invertebrate groups, are also evident, but no attempt was made to synthesize the extent and causes of these on a

Table 1.2 Summary of numbers of major invertebrate taxa included in *The IUCN invertebrate red data book* (Wells *et al.* 1983)

Taxon group	Number of species/ subspecies	Main areas of world/foci
Cnidaria		
Anthozoa	9	Mediterranean/Pacific/Caribbean: mostly corals
Platyhelminthes	2	North American planarians
Nemertinea	8	Representatives of 12 terrestrial taxa, various localities
Mollusca		
Gastropoda	88	Widespread. High proportion of Mexican freshwater snails (12), North American freshwater mussels (26), Hawaiian terrestrial snails (41) and Moorean tree snails (11)
Bivalvia	35	
Annelida		
Hirudinea	1	Europe: medicinal leech
Oligochaeta	[98]*	Mainly, South African acanthodrilines (90) and various 'giant earthworms'
Arthropoda		
Merostomata	4	Horseshoe crabs: only living taxa
Arachnida	7	Spiders (6); harvestmen (1)
Crustacea	15	Tasmanian anaspidids (5), various crayfish, coconut crab
Insecta	[71]*	13 orders included, see p. 00
Onychophora	[All]*	South temperate zone and tropics
Echinodermata	2	European sea urchins

[ ]*denotes inclusion of general/families without precise numbers.
Note: Other taxa, such as Porifera, are also included in *The IUCN invertebrate red data book*, but without species-level data sheets.

global basis until the appearance of *The IUCN invertebrate red data book* (Wells *et al.* 1983) (p. 120), a milestone in the development of invertebrate conservation. Representatives of many phyla are noted there (Table 1.2). Despite this diversity of concerns, a far greater number of declining taxa are *not* included, but that volume publicized the needs of invertebrates in a forceful and constructive way for the first time. It paved the way for many further listings of taxa considered to be vulnerable or threatened, and helped to focus attention on the numerous threats to invertebrates.

Confronting fully the needs of invertebrate conservation is difficult, but three major dimensions recur. These are science, logistics and ethics (Fig. 1.2). 'Science' reflects knowledge, its interpretation, and how to apply it most effectively to conservation. 'Logistics' determines the rate and extent of acquisition of additional knowledge, and the effectiveness and capability of applying the science in practical contexts. The underlying considerations of 'ethics', posing questions of almost imponderable difficulty over how priorities for action may be set, determines strictures and attitudes, including political and public sympathy for conservation action in relation to a suite of different 'values' (Chapter 2).

The complexity of responses needed is exacerbated by un-evenness of knowledge, diversity distribution, and applicability of common ecological concepts. A major difference between terrestrial and marine ecosystems, for example,

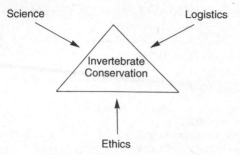

Fig. 1.2 The dimensions of invertebrate conservation (after New 1993*b*).

is the *relative* lack of knowledge of marine organismal diversity. Whereas conservation concern on land extends over the whole range of terrestrial biomes, the most tangible need for marine systems is for the relatively small component of coastal environments—such as coastal wetlands, lagoons, coral reefs, and mangroves, where human impacts are visible and often massive. Trends in the major oceans and the profundal areas are not generally as easy to appraise and, although threatened by pollution in various ways, any form of species-oriented or taxon-focused conservation is of much smaller concern there than on land. As di Castri *et al.* (1992) comment, a different 'thinking' is needed for marine systems than for terrestrial ones, and habitat protection of coastal zones is the most obvious major need for marine diversity conservation. One important aspect of this is that some workers claim that the concept of 'community' as a discrete biotic assemblage associated with a particular habitat, used widely for terrestrial systems, does not transfer easily to the marine environment because of the much more widespread nature of many marine invertebrates, and prefer a more neutral term such as 'assemblage' for, for example, local benthic faunas. Physical factors, such as salinity, temperature, and the nature of the substrate, may be more important than biological factors (other species) in determining distribution of many marine species, so that attempts to define 'communities' based on biological values (such as dominant species) may lack biological reality in some instances.

The problems of understanding invertebrate biodiversity are unlikely to be overcome fully in the foreseeable future. Yet their predominance and key roles in most natural ecosystems collectively render them vital in assuring the well-being of these systems. The themes of 'invertebrate importance' and how they may be utilized to indicate environmental health are developed in the next chapter.

FURTHER READING

Bond, W. J. (1989). Describing and conserving biotic diversity. In *Biotic diversity in southern Africa* ed. B.J. Huntley, pp. 2–18. Oxford University Press, Cape Town.

Ehrenfeld, D. W. (1970). *Biological conservation*. Holt Rinehardt and Winston, New York.

Groombridge, B.C. (1992). *Global biodiversity*. World Conservation Monitoring Centre, Cambridge.

McNeely, J. A., Miller, K. R., Reid, W. V., Mittermaier, R. A., and Werner, T. B. (1989). *Conserving the world's biological diversity*. IUCN, Gland.

Parker, S. P. (1982). *Synopsis and classification of living organisms*. McGraw-Hill, New York.

Samways, M. J. (1994). *Insect conservation biology*. Chapman & Hall, London.

Soulé, M. E. (ed.) (1986). *Conservation biology, the science of scarcity and diversity*. Sinauer Associates, Massachusetts.

Wilson, E. O. and Peter, F. M. (ed.) (1988). *Biodiversity*. National Academy Press, Washington, DC.

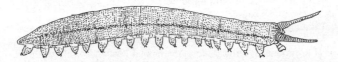

2 WHY CONSERVE INVERTEBRATES?

INTRODUCTION

Conservation of invertebrates is not necessarily incompatible with killing them. Many people pass their working lives trying to kill invertebrates as efficiently as possible, and in some instances this practice accompanies a widespread accepted need for conservation. This may be to assure the sustainability of food stocks of edible molluscs and crustaceans in the face of intensive harvesting or exploitation, for example, or to maintain rare butterflies which can be harvested in small numbers for sale to collectors. But in other cases conservation is not usually considered. The pest-control industry is large and vital in protecting human crops and other commodities from the depredations of insects and other pests, and in maintaining the health of domestic stock by eliminating disease and disease vectors. It has helped to foster the widespread public (and political) image that 'bugs' should be killed indiscriminately. This is a real obstacle to gaining sympathy for invertebrate conservation, and is clearly incompatible with broad advocacy of conservation. Overcoming public prejudice demands education at all levels to counter this unfavourable general image of invertebrates. It is necessary to clarify and stress their numerous positive values, and to emphasize that those which are regarded as pests—even though they may be very important in particular contexts—are indeed a very small minority of the species inhabiting our world. Without public and political sympathy, based on values which can be readily appreciated by people, it becomes extremely difficult to make effective cases for conservation, especially where this involves resolving conflicts over land use in order to guarantee reserved habitat for an invertebrate.

As in other conservation contexts, the issues range from pragmatic and tangible to ethical and idealistic, with different sectors of the human population emphasizing different priorities in this broad spectrum. Collectively, for invertebrates, positive values for conservation are overwhelming and the main task is to formulate, promote, and convey these effectively to people whose decisions may affect the animals' future. Reliance on pragmatic or anthropocentric values may lead to neglect of true ethical values (Lockwood 1987), but this priority does reflect political reality in the 'use it or lose it' school of conservation assessment. One could argue, though, that general sentiments of 'species rights' and 'environmental sensitivity' are integral components of any worthy conservation programme, at least at some level of execution. The four main broad categories of values, defined on p. 22, are (1) commodity, (2) amenity, (3) moral, and (4) functional or ecological, and all are used extensively in assigning priorities (p. 22). Pragmatic reasons for conserving invertebrates encompass their central integrating roles in ecosystems, so that without them most natural biotic systems would not function properly (or at all), and their positive direct benefits to people. Slightly less tangible values are exemplified by the importance of invertebrates as study vehicles and educational tools, and their use as ecological monitors (p. 25).

The recent *European charter for invertebrates* (European Community 1986) sets out clearly many of the compelling positive values of invertebrates,

Table 2.1 Main points of the *European charter for invertebrates* (European Community 1986)

---

1. Invertebrates are the most important component of wild fauna, both in number of species and biomass
2. Invertebrates are an important source of food for animals
3. Invertebrates may also constitute a source of food for mankind
4. Invertebrates are vital to the fertility and to the fertilization and production of the vast majority of cultivated plants
5. Invertebrates are useful in protecting farming, forestry, animal husbandry, human health, and water purity
6. Invertebrates are valuable aids for medicine, industry, and crafts
7. Many invertebrates are of great aesthetic value
8. Some invertebrates may harm human activities but their populations may be controlled naturally by other invertebrates
9. Mankind can benefit greatly from enhanced knowledge of invertebrates
10. Terrestrial, aquatic, and aerial invertebrates should be protected from possible causes of damage, impairment, or destruction

---

and these are recapitulated below. That document has been the impetus and template for several other, similar, regionally based statements.

The main points of the *European charter* (Table 2.1) are all self-evident, important, and relevant, but some merit brief discussion here, to emphasize some of the general themes.

ECOLOGICAL IMPORTANCE

'Ecological importance' is the most vital reason for maintaining biodiversity, and necessitates the successful maintenance of ecosystem functions, rather than simply that of individual component species, coupled with the realization that significant loss of species will change or disrupt the ecosystem of which they were a part. *If* the ecosystem functions can indeed be maintained, loss of species may in any case be minimized. Practical approaches to this tend to emphasize preventing decline in species richness, but understanding the intricate relationships between biodiversity and ecosystem function is critical to understanding how to manage (p. 71) habitats to conserve the greatest number of taxa. Walker's (1992) elegant analogy, which emphasizes the differences in ecological roles between species in a community, is that some are equivalent to 'drivers' of the system and others are 'passengers'. The former are akin to keystone taxa (p. 23), and the loss of these important species may cause major perturbation or change to the communities. Loss

of passengers, in contrast, may make little difference. The latter are, in this sense, ecologically redundant and are 'expendable' in terms of maintaining the systems of which they are a part (if not for their intrinsic interest or worth). This assumption then leads to attempts to determine what (if any) 'ecological redundancy' exists in such complex biological systems, and to identify the redundant components as possibly having a lowered conservation priority. In turn, this may lead toward pragmatic evaluation of a decline in diversity. This approach is *not*, in any way, equivalent to *advocating* removal of species or their sacrifice. In advancing a rational approach to biodiversity conservation, Walker (1992) advocated strongly the wisdom of concentrating attention on the components and aspects of biodiversity that are critical to maintaining the resilience of ecosystems, and this approach has been implied by a number of invertebrate biologists seeking rational priorities for conservation focus. The high diversity and biomass, and the ecological predominance, of invertebrates have already been stressed. They play predominant roles in most ecosystem processes and are necessary links of food webs in any community. All organisms, living or dead, are eaten or otherwise used by invertebrates. Many organisms depend entirely on particular invertebrates for their own sustained existence: many flowering plants for pollination, for example. Even at the more 'domestic' level, most cultivated plants are pollinated by insects, and enhancement of pollination levels of orchard and field crops by shifting bee-keeping to coincide with crop flowering seasons is itself a major aspect of the bee-keeping industry in improving yields and productivity of many food and textile crops.

The role of earthworms, and other invertebrates, in soil aeration and litter decomposition is instrumental in fostering vegetation growth, and exemplifies the less obvious ecological roles on which other organisms depend. Some European earthworms have been imported to Australia for the express purpose of improving grassland productivity and, in turn, improving cattle and sheep production, for example. In a related topic, avoidance of pasture staling by the dung of introduced cattle and horses in Australia, through its long persistence and unsuitability for consumption by native insects, has been obviated largely by importation of dung beetles from Africa and Europe, which rapidly break down the dung. They thereby ensure efficient recycling of the dung materials, continuation of pasture productivity, and also remove the main breeding sites of nuisance flies. Soil structure can be improved in many ways by invertebrates, and these are associated with three major activities, namely burrowing (changing porosity and improving aeration and water passage), humification (improving stability and friability through nutrient recycling), and mixing (changing the soil profile) (Abbott 1990; Hutson 1990). These activities involve many different invertebrate taxa, sometimes in enormous numbers.

Decomposition of plant and animal material is achieved predominantly by invertebrate action in many places and in all major ecosystems. Major ecological structures, the basis for diverse communities, such as coral reefs and atolls, derive directly from invertebrate activity.

As one specific context illustrating diverse ecological worth, Janzen (1987) considered the fundamental, but largely undocumented, roles of insects in tropical forest; similar appraisal could be made for other invertebrate groups and other ecosystems. Janzen noted several (of many possible) ways in which he considered insects as part of the 'glue and building blocks' of dry forest in Santa Rosa, Costa Rica:

1. 'Insect species are not merely interchangeable bits of nutrients.' Many insectivorous vertebrates select, and depend on, a particular suite of prey taxa rather than just taking random prey, and the particular prey spectrum eaten may be highly correlated with the consumer's metabolism and future reproductive performance.

2. Insect seed-killers. Beetles, particularly Bruchidae and Curculionidae in Janzen's study, feed on and destroy a high proportion of legume seeds. If the beetles were eliminated, perhaps by the destruction of adult refugia by agriculture, particular legume species could, in theory, become superabundant and affect the balance of forest diversity. Janzen cited the effect of one weevil, *Apion johnschmiti*, which eats 80 per cent or more of seeds of the tree *Ateleia herbert-smithi* (characteristic of secondary succession), and emphasized the importance of this case to the forest's security from invasion.

3. Pollinators. It is commonly assumed that pollinator species 'removed' from a community are replaced by others, but this is generally rather unlikely because of the highly specialized co-adaptations of the partners in many such associations. Some pollinators, such as many of the hawkmoths (Sphingidae), in Santa Rosa migrate and spend part of the year in other habitats which might also be subject to despoliation. The general point is that such pollinators (and, indeed, other consumers) vital to the well-being of a local community might not themselves be permanent residents in that community.

USE OF INVERTEBRATES AS FOOD

Traditional use of invertebrates and their products as human food is widespread. Some such uses form the basis of substantial industries, some widespread and others of key significance in local economies. In particular, marine (especially) and freshwater molluscs and crustaceans, and terrestrial molluscs, are important in commerce, and many form the basis of major fisheries and farming activities. These range from large-scale exploitation and cultivation to

smaller-scale production or individual farms for
local or domestic use—such as recent trends in
Australia to raise certain freshwater crayfish in
farm dams as a cash crop to offset an ailing
agricultural economy.

Bardach *et al*. (1972) traced the early history
of aquaculture for many groups of invertebrates.
Lobsters and crayfish have long been used by
people as 'primary protein', but their value as
gourmet foods in wealthier parts of the world
has been a massive impetus to aquaculture for
mass production under varying levels of confine-
ment and control. Expansion of such industry is
important in helping to alleviate pressures on
wild populations (p. 102). In Lousiana, the most
important crayfish-producing area of the United
States noted by Bardach *et al*. (1972), more than
800 000 kg of wild-caught crayfish were supple-
mented then by 1.2 million kg produced on the
6000–7000 ha of crayfish farms in the state.

Some shrimp and prawn species have been
cultivated in Asia for five centuries and more,
commonly as a subsidiary activity to fish culture.
The predominant taxa used are members of the
prawn families Penaeidae and Palaemonidae.
Singapore was the first country to concentrate
on shrimp culture in modern times, with introduction of an industry in modified mangrove
swamps in 1937. Considerable advances in technology have been made also in Taiwan, where
it is usual to cultivate several species in the
same ponds (Bardach *et al*. 1972). Considerable improvements have been associated with
increasingly effective international marketing in
recent years.

The best-known terrestrial invertebrate food
industry is bee-keeping, with honey and other
hive products (propolis, wax) being of great economic importance. Consumption of grasshoppers,
termites, and a wide range of other taxa is widespread in many of the less developed parts of
the world, where insects can constitute substantial
proportions of dietary protein. The further potential of insects as human food is explored by Myers
(1983), DeFoliart (1989), and Vane-Wright (1991),
amongst others. They cite many examples, and
there is little doubt of the nutritional values if the
inherent prejudices of many Westerners against
eating insects could be overcome.

BIOLOGICAL CONTROL

Protection of food crops from depredations of
pests, predominantly plant-eating invertebrates,
often involves the deliberate manipulation or
release of 'natural enemies', predators or parasites, to effect biological control. Mass production of these agents, most commonly insects but
also sometimes other arthropods and nematode
worms, is an important aspect of integrated
pest management, and many such agents have
been introduced to parts of the world well
beyond their natural range to combat 'exotic
pests' (p. 55).

This discipline draws on the ecological integration in natural communities, whereby most
species potentially harmful to human endeavour
are controlled naturally and unobtrusively by a
suite of other invertebrates, so that for much of
the time their presence is unnoticed or tolerable.
With establishment of monoculture crops, often
in exotic situations, plants may be reached by
consumers unchecked by their natural enemies,
so that this usual natural balance is destroyed.
'Pest status' is often achieved in this way. Biological control measures can eventually help to
restore a balance in an environmentally acceptable manner, substituting (at least, in part) for
pesticide use.

OTHER 'COMMODITY USES'

A great diversity of other uses of invertebrates
occurs, some of them of considerable antiquity.
Examples are:

1. Sericulture. The use of silkworms, mainly
the domesticated moth *Bombyx mori*, to produce silk was developed by the Chinese many
centuries ago. The moth has been exported to
many parts of the world.

2. Marine sponges, which have long been
harvested for human domestic and industrial
use (p. 143). Some are currently being screened
for cancer-curing drugs, as part of a much
broader awareness of the pharmaceutical potential

of various invertebrate groups. Some pharmaceutical firms are determinedly moving to increase their surveys for products occurring naturally in invertebrates, rather than continuing to modify known chemicals in the laboratory. People already utilize hundreds—perhaps thousands—of products which owe their origins to wild animals or plants, and the vast untapped cornucopia of sources of possible 'miracle drugs' and the like is coming to receive more and more attention. One Caribbean sponge, for example, has compounds which seem to be effective against herpes, encephalitis, and some leukaemias. Generally, the use of invertebrates in medicine is almost entirely unexplored, and some authorities believe that this area is ripe for exploration and exploitation. Myers (1983), for example, noted the possible medical benefits which might accrue from marine life, and emphasized the vast array of biocompounds present in these animals, such as the sponge compounds noted here.

It seems that many invertebrates are sources of potentially exploitable biocompounds different from those of their close relatives, emphasizing further the wisdom of conserving species rather than just representatives of higher taxonomic categories.

3. Dyestuffs produced from molluscs, a practice going back to the Romans and Greeks exploiting Mediterranean sea species (see below).

4. Use of shells and corals for jewellery and ornaments, mother-of-pearl (button manufacture), 'collectable items', and building materials.

5. Local hunting of earthworms, marine worms and molluscs, various insects, tunicates, and others for fishing bait or use as fish food in aquaculture.

6. Earthworm protein, produced through intensive mass-rearing, or vermiculture, is a valuable food supplement for farm animals. This industry is still in its infancy (Sabine 1988), and has as a by-product enriched compost for horticulture.

7. Use of medicinal leeches (p. 133) in treatments to reduce bruising and swelling after surgery.

CULTURAL ASPECTS OF INVERTEBRATES

Aesthetic appreciation has led to some groups (perhaps, particularly, some insects and molluscs) playing significant roles in human culture. Dragonflies have long been important in Japan, as have cicadas in China. The ancient Greeks used the same name ('psyche') for 'butterfly' and 'soul', with resultant mystical appreciation and connotations for these insects (Davies and Kathirithamby 1986). The Chinese had three main periods when crickets were an important part of their culture (Hammond 1983)—in early times they were appreciated for their song; from about AD 618–906 they were kept in cages for enjoyment, and from about AD 960–1278 the sport of cricket fighting was developed strongly. The latter has recently undergone considerable resurgence, so that Yen (1994) quoted estimates that 100 000 crickets are now sold every day in Shanghai alone. The ancient Chinese recognized the cicada as a symbol of the resurrection (paralleling the Greek respect for butterflies) and the praying mantis as representing bravery, but neither of these apparently had the impact of crickets, which formed the basis for elaborate art-forms for cages and adjuncts such as 'ticklers' to incite the insects to sing or fight.

The *European charter for invertebrates* comments: 'invertebrates [are] a major source of inspiration for ordinary people and artists'. Dance (1966) noted that mollusc shells are associated with cultures of pre-Dynastic Egypt, prehistoric Europe, pre-Columbian Ecuador, and prehistoric Mexico; that shell trumpets were important ceremonial implements; that cloth dyed with the rich purple of molluscs such as *Murex* was reserved for royalty, and that cowries were used as legal tender. Perhaps cowries were at the foundation of 'shell-culture', because of their fancied resemblance to human female genitalia which was believed to confer particular powers on the shells. The scallop motif is widespread in classical architecture and ornamentation. The presence of exotic shells in the ruins of Pompeii shows that even shell-collecting may antedate the devastating eruption of Vesuvius in AD 79.

INVERTEBRATES AS RESEARCH TOOLS

Invertebrates have been instrumental in leading to the elucidation of many biological disciplines and principles. The vinegar fly, *Drosophila*, is still an important vehicle for research and teaching in genetics, for example (P. A. Parsons 1983). Knowledge of many aspects of behaviour, physiology, and population biology has been developed mainly from work on particular key invertebrates. As another example, marine molluscs such as *Aplysia* and *Octopus* played significant roles in understanding the functioning of nervous systems. Their full potential, though, has yet to be realized. Recent studies on earthworms, for instance, have shown that they might possess early stages of complex oxygen-carrying proteins which are important in higher organisms. Studies of these 'simpler' systems may lead to greater understanding of their origins and functions (Sabine 1988).

SUSTAINABLE UTILIZATION

Any commercial use of invertebrates depends in the long term on the sustainability of the species (or population) being exploited or harvested. This is most obvious in cases of high human need, such as large-scale fishing of marine species such as squid and prawns.

The value of local enterprises involved with invertebrates can be significant in ways less obvious than the usual forms of direct large-scale exploitation. Importantly, they can help to convey the need for conservation as a basis for sustained use. They may markedly enhance the values of protected areas, or demonstrate the need to protect additional sites or habitats, in the eyes of local people who otherwise might react antagonistically to loss of their traditional lands by exclusion or change. Two examples are noted below.

1. The first involves bee-keeping and gathering of saturniid moth caterpillars ('mopani worms') for food in Malawi (Munthali and Mughogho 1992), being an integrating factor between local rural people and biological conservation. High caterpillar yields were obtained from vegetation 1–3 m high. Rotational burning and coppicing to promote this vegetation form would also promote sustainable utilization, facilitating good harvesting through ease of access, and reduce (illegal) cutting of trees to obtain caterpillars. As elsewhere, access of bees to the nectar supplies of protected areas may be important as nearby vegetation is removed progressively (cf. p. 64).

A survey (40 respondents) indicated that all families when living in the Kasunga National Park practised bee-keeping and gathered caterpillars, but after exclusion from the Park only a third kept bees and none harvested mopani worms. Cash derived from these resources within the National Park was seen as important in wildlife management in Malawi by fostering appreciation of the need for management and gaining the support of the local rural community (Munthali and Mughogho 1992).

2. Another practice leading to considerable benefit to rural people is ranching or farming of butterflies, whereby rational harvesting of butterfly populations to satisfy the demands of collectors and supply livestock for display in 'butterfly houses' (p. 67) is undertaken from areas planted with caterpillar foodplants and adult nectar sources. Now undertaken in several parts of the world, this practice was pioneered in Papua New Guinea (Clark and Landford 1991) through building up a nationwide network of local suppliers. An instructional 'farming manual' (Parsons 1978) led progressively to a widespread rural industry centred on the government-founded Insect Farming and Trading Agency (IFTA), now part of the Papua New Guinea University of Technology's commercial enterprises, at Bulolo. The aim of the IFTA is to return as much income as possible to the farmers and collectors who provide the insects for international sale. Emphasis has been on providing an income based on a sustainable resource (and

the needs of sustainability are emphasized) for people in areas which may not be amenable to crop or livestock production. More than 500 farmers/collectors, from 14/19 Provinces of Papua New Guinea are now involved, and this number seems poised for substantial increase.

In some places, butterfly populations have actually been increased by the operation, because of the increased supply of normally sparse foodplants. The industry is viewed as viable, and one which is sufficiently lucrative to deter forest clearing for agriculture as an alternative source of income. Butterfly farming (technically, ranching: p. 102) occurs in harmony with rainforest environments and thus serves an important 'umbrella' conservation role (National Research Council 1983; Clark and Landford 1991; Orsak 1993). Much original scientific information relevant to conservation has been gathered and collated through the operations of IFTA.

Both the above examples demonstrate the importance of incorporating the needs of local people into 'conservation through development', a strategy which is coming to be of greater importance over much of the tropics.

EDUCATION AND ATTITUDES

Despite the positive values noted above for invertebrates, public perception of them is all too often that of 'creepy-crawlies', variously the subjects of fear, loathing, or mistrust (Speight 1986a). This attitude is fostered strongly by many Western news media, which tend to report enthusiastically on venomous spiders, jellyfish, Hymenoptera ('killer bees'), and the like, while largely ignoring less sensational positive values (Yen 1994), although this attitude is gradually changing. The problem of public education to restore the balance is important in fostering positive attitudes towards invertebrates and obviating their widespread perception as threats or 'enemies', and this aspect is discussed in Chapter 8. It is important, though, to acknowledge that many normally harmless and innocuous invertebrates, not just those that can bite or sting, suffer from such unpopular images. In her

surveys of a giant earthworm in Australia, van Praagh (1994) found the spectacular newspaper headline 'Giant 13-foot worm strangles baby': the worm killed the infant in 'a hideous and deliberate attack before the mother's eyes'. The public feelings engendered by reports such as this are extraordinarily difficult to dispel.

The 'values' noted above counter clearly the common public comment about invertebrate conservation: 'why bother?'. Once ethical parameters are also incorporated the case becomes overwhelming. However, the sheer magnitude of the task also fosters doubts from many people, couched in terms such as: 'if there are so many species, and we know so little about them, what do they matter anyway?' The proprieties and problems involved in 'ranking' species for conservation preference, in essence evaluating and comparing 'worth', are formidable (New 1991b) (p. 75). This difficulty recurs in conservation planning governed by the need to determine optimal use of very restricted resources, but much of the current thrust of invertebrate conservation involves a rather different form of setting priorities: that of detecting the taxa of greatest use as tools in conservation assessment and utilizing these in the broadest possible contexts of ecological conservation. Evaluation of ecological worth and responses to particular kinds of ecological perturbation thus become important, in this duality of taxa as 'subjects' and 'tools' in practical conservation.

Definition of functional roles (guilds, often equated to their place in food webs) of invertebrates focuses attention on some aspects of ecological worth, but there are clearly problems in pursuing this approach fully. Walker (1992) suggested a four-step sequence, as follows:

1. Guild analysis of the community, with attempts to subdivide the species of each guild on the basis of their functional attributes.

2. Determining the number of species in each guild (in general, the group of taxa with a similar trophic role or feeding habit). Guilds represented by few (or, even, a single) species are unlikely to be able to withstand loss of taxa and are an urgent focus for conservation.

3. Examining in more detail the interaction between species in each guild, especially to determine whether density compensation (increase in the abundance of the species which remain) occurs with removal of species.

4. Considering the relative importance of the functional groups in terms of their abundance influencing processes in the community and ecosystem, and especially in relation to system stability.

Completion of this approach is obviously extremely difficult for most invertebrates, because we lack knowledge of the precise guild attributes of most of the taxa, and such 'functional approaches' need to be combined effectively with taxonomic approaches in setting conservation priorities. As Cranston (1990) emphasized, approximations in guild analysis for invertebrates tend to accumulate in two ways: from lack of species recognition, so that identification to family or genus level may lump together species with very different biology, and from allocating species whose habits are unknown to the guilds documented for their better-known relatives.

The two are complementary in designing overall strategies for conservation success:

(1) the functional group approach gives major attention to taxa which are of greatest importance in helping to minimize loss of biodiversity, whereas

(2) taxonomic approaches are relevant in assuring that maximum biodiversity is included in habitat patches recommended as reserves (Chapter 4) and ensuring that reserves complement each other in total representation of biota.

These, and other approaches to setting priorities, need much further discussion to help the scientific community agree on worthwhile protocols for invertebrate conservation, but indicate avenues likely to help us to assess priorities objectively according to a range of different criteria. Other criteria, from the range of 'values' noted earlier, are more difficult to appraise scientifically because they include components of subjectivity or variability. Norton's (1986) species-based values of 'commodity', 'amenity', and 'moral' (below) can lead to subjective, even highly emotive, discussion, particularly of the last two categories (Regan 1986; other papers in Norton 1986; Lockwood 1987; New 1991b; for examples and included references). 'Commodity value' is persuasive politically and refers directly to a species' role in the market place (Myers 1983), with attendant variables of fluctuating supply and demand. A species has commodity value if it can be sold or otherwise generate income. 'Amenity value' *sensu* Norton is adduced if a species improves the quality of peoples' lives in a non-material way. Many people like to see butterflies, or a diversity of sea-shore or reef life, for example, and recreational pursuits and the tourist industry can have increased markets because of the presence of 'charismatic' or 'pleasing' species, and reduced markets if these are absent. 'Moral values' are an intrinsic part of conservation evaluation to many people, but the ethical implications of species loss, and of ranking species in some way for preferential treatment may impinge directly on other values. Regan's (1986) comment to the effect that, in the absence of commodity or ecological values, whether people see a duty to conserve a species may depend simply on if they 'like' it, expresses the levels of subjectivity involved, and has serious implications for invertebrates because of the high proportion of species which, traditionally, people do *not* like, or with which they are entirely unfamiliar.

Other aspects of priority taxon selection are noted in Chapter 4. The remainder of this chapter discusses some of the ecological values of invertebrates and how these may be applied in practical conservation.

ECOLOGICAL PRIORITIES FOR INVERTEBRATE CONSERVATION

Within the framework of ecological priorities noted earlier, four kinds of invertebrates have been discussed as priorities, irrespective of their taxonomic status. These have elements in common, but each category has features peculiar to it. Collectively

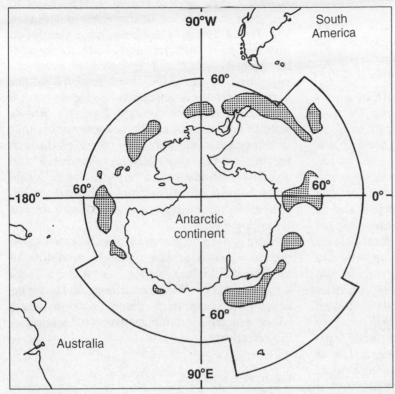

Fig. 2.1 Distribution of krill: major concentrations of *Euphausia* in the Antarctic are shaded (after Knox 1984). The heavy boundary line delimits the area of application of the Convention on the Conservation of Antarctic Marine Living Resources.

they represent a range of groups that can facilitate valid short-cuts to biodiversity conservation, by being amenable to collection, or helping to make people more aware of invertebrates and their importance in the natural world.

These groups are:

(1) keystone species;

(2) umbrella species;

(3) flagship species or groups;

(4) indicator species or groups.

Keystone species

Defined as those on which the local community or assemblage functionally depends, these are arguably the most important invertebrates in any ecosystem. The concept of any species having such critical ecological importance is queried by many ecologists, but is difficult to refute entirely. As the name suggests, if keystone taxa are harmed the effects of their loss may ramify through the whole assemblage or community rapidly and irreparably. The concept is more typically defined by vertebrate examples, but representative invertebrate keystone taxa may include abundant species such as the predominant corals of a local reef. A clearly definable marine example is 'krill' in Antarctic ocean waters. Although krill are not in need of conservation at present, they illustrate well the role of a putative keystone invertebrate.

Krill, a group of euphausid crustaceans, the most predominant of which is *Euphausia superba*, though around 11 other species are also represented, occur in a circumpolar band around Antarctica (Fig. 2.1) with locally dense concentrations probably maintained by local water currents. Three different kinds of aggregation have been delimited (Cram *et al.* 1979):

(1) 'layers', thin or thick vertically, can extend for many kilometres: one of more than 200 km was recorded by these authors;

(2) 'swarms', aggregations of variable thickness

but horizontally extending for tens to hundreds of metres and

(3) 'superswarms', up to a kilometre or more in diameter with very high densities; one such superswarm was estimated to contain between 5 and 10 million tonnes of krill.

Krill feed mainly on diatoms, and are an important food for vertebrates, especially for penguins, seals, and whales. In contrast to those in the other oceans, the major food chains in the Antarctic are short and simple, of the form phytoplankton–krill–vertebrate as a unifying base. Krill therefore play a major role in converting plant to animal biomass and in constituting about half the standing crop of zooplankton (Everson 1981). As the dominant herbivore, *E. superba* affects the growth and survival of the major vertebrate taxa of the Antarctic, as well as of squid. Vast quantities of krill are fundamental to continued functioning of Antarctic ecosystems (Knox 1984).

Fishing for krill for use as stock food, fertilizer, or human consumption, has been seen as a major opportunity for commercial enterprise, with a tendency to switch harvesting activities in the Antarctic from whales (the top trophic level) to lower ones (squid as an intermediate one, krill) because of widespread public and political concern for whales and the increasing scarcity of some species. Many people have warned that overexploitation of krill could have dramatic ramifications throughout the whole ecosystem, although it has been implied that up to 100 million tonnes a year could be harvested sustainably. Estimation of future yields is rendered difficult by krill's non-uniform distribution, in aggregations as noted above.

Krill fit well the role of 'foundation species', defined by Dayton (1972) as species at low levels in a food web which contribute in a major way to community structure. Understanding its biology and population dynamics is a key theme in Antarctic conservation. Capture quotas are set, currently recommended at 1.5 million tonnes/year, around four times the actual level of exploitation.

As a possible terrestrial example, also one not an immediate target for conservation, the abundant and diverse termite fauna of arid Australia may be a keystone group linked to the great diversity of lizards present (Morton and James 1988). They may have developed because poor soil fertility, with consequent nutritionally poor and sporadically produced vegetation, favoured their well-being over other herbivore groups which depend on higher-quality food. The termites provide food for lizards directly and also indirectly, because some other invertebrate groups may have increased through preying on termites. Termite abundance and diversity declines from infertile to more fertile areas, coincident with declines in lizard diversity. Similar situations may occur in African deserts (Pianka 1989).

More generally, 'keystone' species can include predators, such as large mobile predators in the intertidal zone, which can have a large range of direct and indirect effects on the members of the community (Paine 1974), or others which influence community structure by harvesting selectively.

Umbrella species

'Notable' species (often with many characteristics of flagship taxa, below) characteristic of a particular community or habitat, whose safety can assure (or help to assure) that of many less conspicuous or less well-known taxa in the places where they live. They need not play any major integrating role in the functioning of the community and differ thus from keystone species.

As one example, velvet worms (Onychophora) in humid forest habitats such as rotting logs, in parts of the tropics and southern temperate regions, and in caves are among the most notable and conspicuous members of the specialized communities which depend on those habitats. Their very presence is sufficient to demonstrate an unusual community, and protection of habitats or sites that is motivated and facilitated by the incidence of Onychophora can do much to protect the multitude of other taxa living in wet forest litter or in caves by reducing the incidence of threats and despoliation of their environments (New 1995).

Flagship species or groups

These serve to increase awareness of conservation need by helping to gain public and political sympathy, based on their appeal to people. Particular species can be involved, but whole groups of 'charismatic' invertebrates may also constitute flagship taxa.

The paramount invertebrate group utilized so far as flagships are the butterflies, and Collins and Morris (1985) and New and Collins (1991) sought to promote awareness for swallowtails, birdwings, and related forms (constituting the family Papilionidae) in this way, as well as their more practical uses in faunal assessment and as indicators (below). The essential feature for flagships is popular, usually aesthetic, appeal so that the common prejudice against them as invertebrates is absent, and interest and sympathy for them can be fostered.

The following features are useful for promoting flagship groups of invertebrates, and several of these characteristics overlap with those needed for indicator taxa (p. 27).

1. Taxonomy well known, with many (most) species easy to recognize; if this can be done without capture it is an advantage in survey and regional/site assessment. Swallowtails, for example, can often be identified to species by using binoculars or even just the naked eye, as they are predominantly large, colourful, and diurnal. This is not the case for most invertebrate groups.

2. Ability to engender public sympathy for their well-being, either based on aesthetic value, or commodity value, or both of these. Very few invertebrate groups are 'charismatic' in the way that most mammals and birds appeal instantly to people. However, a range of relatively unlikely taxa have been variously adopted as 'local emblems' and the like: the giant Gippsland earthworm (*Megascolides australis*) (p. 137) in part of south-eastern Australia, for example, has given its name to a local agricultural festival.

3. The groups should be relatively diverse and widespread, but with localized or narrowly endemic taxa which can be used to monitor local community health and to foster local 'pride' or goodwill as part of a broader conservation awareness.

4. They should frequent an array of different habitats, and contain specialist species which respond to habitat change with those responses, at least in broad terms, definable and predictable. They can thus combine characters of indicators and umbrella species effectively.

The main value of flagship taxa is their facility to influence public policy. However much sound biological information is available on a group of animals or plants, it is redundant in effective conservation management unless it can be communicated positively and help to engender change or rational response. 'Flagships' help to gain that vital enhancement that people value the organisms concerned, want them to thrive, and will support moves to ensure this—even at some other cost. This may be because of perceived or actual financial benefit (ranching or farming butterflies, p. 102), more indirect community financial benefit (such as through increased tourism: for example to the overwintering sites of the monarch butterfly, *Danaus plexippus* in California and Mexico), or because the animals are seen in themselves as symbols of a healthy natural environment.

Many flagship species are, in essence, 'emblems' for local or national conservation efforts, and can have important direct benefits to other taxa, as well as the more intangible benefits to their natural communities. Whalley (1989), in discussing the re-introduction of the large blue butterfly (*Maculinea arion*) to Britain (p. 138) noted that, directly as a result of publicity for this species, substantial amounts of money were raised also for *other* conservation projects on butterflies.

Indicator species

In the broadest sense, any species of animal or plant 'indicates' a particular suite of environmental conditions, those best fitted to its own way of life and demands. The major role

Table 2.2 Criteria used to select organisms used for toxicity testing in freshwater ecosystems (after Buikema *et al.* 1982)

1.	The organism should represent an ecologically important group (criteria such as taxonomy, trophic level, ecological role)
2.	The organism occupies a position within a food chain/web linking it with other important species
3.	The organism is widely available, is amenable to laboratory testing, easily maintained, and genetically stable, so that uniform populations can be tested
4.	There is adequate background data on the organism and its place in the community
5.	The organism should show a consistent response to the toxic substance
6.	The organism should not be prone to disease, excessive levels of parasitism or physical damage

of invertebrates in conservation assessment is to indicate levels of disturbance or change, by decline in the diversity of specialized species, increase in abundance of other taxa, or, more generally, some change in faunal composition from the undisturbed state. Particular invertebrate taxa are particularly sensitive to definable change and can be used to monitor the health of the environments in which they occur by providing a 'mirror' of the well-being of less conspicuous biota. Such groups can be promoted for conservation on the grounds of providing sensitive early warning of the effects of human activities, and a broad background is given here to indicate the diversity of their roles.

The most frequent application is to indicate pollution in various ways, and there are about five major contexts in which this is done (Spellerberg 1993), reflecting different kinds of indicator species:

1. Sentinels—species introduced to an environment as 'early warning devices' or to determine the effect of a pollutant.

2. Detectors—species occurring naturally in an area and which show a measurable response to environmental change.

3. Exploiters—species whose presence indicates disturbance or pollution, and which can then become abundant because their competitors have been eliminated. These are therefore pollution-tolerant taxa, rarely of concern for conservation in their own right.

4. Accumulators—organisms which take up and accumulate chemicals from their environment in measurable quantities.

5. Bioassay organisms—those used in laboratory tests to detect pollutants or to rank levels of toxicity.

Collectively, many different groups of invertebrates can be involved in these activities.

The last of these draws heavily on aquatic invertebrates for toxicity testing in aquatic systems, because many are far more sensitive indicators than fish, formerly used simply because more information was available on them. Criteria for determining which organisms should be used in toxicity testing have been designed by the United States Environmental Protection Agency (Buikema *et al.* 1982) (Table 2.2), and the various parameters that may be used to detect a response emphasize the need for very careful interpretation (Fig. 2.2) (McCahon and Pascoe 1990). Inevitably, the topic of indicators is linked broadly with the major threat of 'pollution' (p. 47) and the discussion of that links with this section. However, the full value of indicator taxa extends also to other aspects of environmental or habitat change.

'Sentinel indicators', or invertebrate equivalents to the miner's canary used to provide early warning of danger, are not common, although the use of honey-bees for this has been promoted. Hives can be shifted easily into an area where contamination by radionuclides, arsenic, cadmium, or other substances may be a concern. Pollutants can be detected by analysing bee products (namely honey, wax, propolis, or transported pollen) or the insects themselves. Use of bees as monitors has been suggested for toxic waste sites (Root 1990), or as an experimental network of hives in Italy to monitor a site by the Po River where a nuclear power station was proposed (Ravetto *et al.*. 1987). Following the Chernobyl accident in 1986, honey-bees in several states of the US were analysed to determine possible radioactive contamination by the isotopes ^{134}Cs and ^{137}Cs. Although levels of radioactivity tended to be higher in west-coast bee samples than in those further east, the levels were not significant toxicologically (Ford *et al.* 1988).

Characteristics and uses of indicators

Resident species are used as indicators in freshwater communities, to monitor aquatic pollution, and the ideal features of such organisms were summarized by Johnson *et al.* (1993) as follows:

1. Individuals should show the same simple correlation between their pollutant content and the average environmental pollutant concentration at all locations and under all conditions.

2. Individuals should not be killed or rendered incapable of reproduction by the maximum levels of the pollutant encountered.

3. The species should be sedentary, so that findings relate directly to the environment in which it occurs.

4. The species should be sufficiently large, or individuals abundant, to provide sufficient tissue for analysis.

5. Individuals should be sufficiently widespread to facilitate comparative assessment in different areas.

6. The species should be sufficiently long-lived to enable sampling of several year classes to provide information on long-term effects.

7. Individuals should be easy to collect.

8. Individuals should be hardy enough to survive handling, if required.

The long-term research use of marine invertebrates as pollution monitors is exemplified by the US 'Mussel Watch' programme, used to monitor a range of chemicals, and commenced in 1976 (Lowman 1979). Two species of mussel (*Mytilus edulis, M. californianus*) and two of oyster (*Ostrea equistris, Crassostrea virginica*) were collected, in annual samples of 25 kg, from each of 107 coastal stations, and the results have emphasized the value of bivalve molluscs as indicators of levels of pollutants introduced at very low concentrations and which can be analysed only after being concentrated through accumulation in animal tissues.

Candidate invertebrates used to monitor pollution may be selected on a number of criteria, discussed by Lower and Kendall (1990):

1. A wide geographical and/or ecological distribution may allow for comparison of trends or data from the same species in different locations.

2. Restricted 'home range' means that the source of pollution may be nearby, rather than far distant—for example, for migratory species.

3. Its presence at an important locality, even if it is not widely distributed elsewhere, may be an important source of information about that locality.

4. Availability of one or more easily measured and/or significant biological endpoint(s) for bioassay in the species.

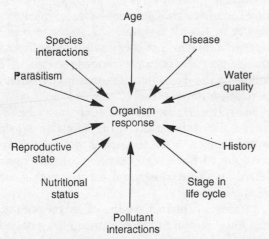

Fig. 2.2 The complexity of biomonitoring. An example of the factors that may affect the response of an insect in testing for toxicity (after McCahon and Pascoe 1990).

5. Ability to use the species to 'probe' a particular aspect of the environment; for example, a particular habitat or trophic level.

6. The ease of accurate recognition and identification of the species, especially when it is to be used by people who are not experts in its taxonomy.

Particular groups of invertebrates may be of much more focused use in conservation assessment than others, because they can be used in specific contexts to monitor environmental changes, to indicate the presence of a given combination of ecological factors, or to otherwise reflect the health of the communities of which they are a part, by responding to some form of disturbance by changes in their abundance or appearance. There is increasing interest in the use of invertebrates as 'indicators' in these ways, so that trends in some of the more diverse and better-documented groups are now widely used to infer other features of aquatic or terrestrial community well-being and to estimate need for remedial management of a habitat or site. The contexts range from general indices of community maturity (nematodes might be useful in assessing soil quality: Bongers 1990) to detecting or monitoring specific chemical or biotic changes.

Invertebrates are thus of considerable importance in environmental impact assessment, thereby indicating needs for threat abatement or habitat restoration, although they have traditionally been neglected unduly, as Rosenberg *et al.* (1986) emphasized for insects. This is due to a wide range of factors, including lack of guidelines over what is required, lack of 'early' species-level identification for most groups, large volumes of material, for which sorting is costly, and small inconspicuous form, in comparison with vertebrates.

However, the most important of these, lack of guidelines, has been overcome partially. Lehmkuhl *et al.* (1984) stressed the need to differentiate between scientific and political processes in setting plans for using insects in conservation work, and that the aims should be clarified carefully. Rosenberg *et al.* (1986) argued that insects

(in their discussion, but other invertebrates also) are ecologically pervasive, that perturbations to the environment impinge on their numerous collective roles, and that they may respond in characteristic ways which are useful in assessing the environmental impact of those perturbations. Particular responses may be of value in many different contexts.

Use of single species as indicators, although well-established in such disciplines as toxicology and pollution control, is open to criticism in some contexts, because these may not predict responses at higher taxonomic or community levels adequately (Cairns 1983). Noss (1990) noted four points that should be emphasized when seeking indicators in any given context:

(1) what is being monitored or assessed, and why;

(2) selection of indicators depends on asking or formulating specific questions which need to be assessed by the monitoring process;

(3) in seeking to monitor one level of ecological operation, indicators might be selected for others; and

(4) many indicators will be specific to given ecosystems, so that general categories of indicators may cut across different ecosystems.

Morphological change
Environmental change is therefore manifest most clearly by loss of sensitive species or change in relative abundance of species, but other kinds of change may also occur in response to stress. The following are examples of structural deformities in relation to pollution effects.

Many species of midges (Diptera: Chironomidae) are very tolerant of pollution, so that water bodies with industrial pollutants may continue to have high numbers of chironomid larvae to the exclusion of most other aquatic invertebrates.

In samples near a source of metal pollution in Ohio, chironomids comprised 86 per cent of all insects, with the proportion decreasing progressively downstream as pollution levels also declined (Winner *et al.* 1980). However, some species are extraordinarily sensitive to changes in water quality and respond to metal pollution (for example) by developing abnormalities of the mouthparts, so that the ligula or labrum, normally bilaterally symmetrical, may be markedly distorted (Fig. 2.3). Such deformities have been linked with a wide range of industrial pollutants, including metals, organochlorine pesticides, radioactive wastes, and coal wastes (Pettigrove 1989). Pettigrove examined the incidence of mouthpart (ligula) deformity in larvae of a midge, *Procladius paludicola,* in the Murray and Darling Rivers, Australia. Most deformities were relatively small, but occasional gross distortions (Fig. 2.3) occurred. In this instance no clear link with any particular pollutant was evident. Even though no pollution may be obvious, the incidence of deformities indicates that chemical imbalance or stress is indeed present at significant levels. Incidence of deformity can be very high: for example, more than 80 per cent of *Chironomus* larvae from the Port Hope Harbour, Ontario, were deformed (Warwick *et al.* 1987) one of the highest levels recorded.

Further examples of similar deformities involve other head appendages of chironomids, the head appendages and cerci of stonefly larvae (Donald 1980), and the tracheal gills of stoneflies and caddis fly larvae (Simpson 1980). In a slightly less obvious effect, the nets which hydropsychid caddis larvae spin to filter food from the water may have an irregular or abnormal mesh, with the frequency of abnormalities decreasing with distance from pollution (Petersen and Petersen 1983).

Developmental distortion can occur in organisms to constitute 'fluctuating asymmetry' (FA), as a more general response to environmental stress. This can be measured accurately, especially in arthropods, as a quantitative difference between the left and right sides of hard exoskeletal structures otherwise not subject to distortion, with the frequency of differences between the two sides being an index of stress (Clarke 1992, 1993).

'Fluctuating asymmetry' may be defined as the random differences that occur between the two sides of a normally bilaterally symmetrical organism, and the rationale of its use as an indicator of stress is that some form of environmental and/or genetic stress during the organism's development may affect developmental

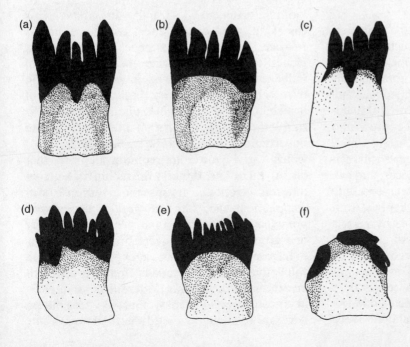

Fig. 2.3 Mouthpart (ligula) deformities in larvae of the midge *Procladius paludicola* from the Murray and Darling Rivers, Australia. (a) Normal ligula; (b–f) various deformities (after Pettigrove 1989).

pathways and lead to an increased level of FA. Such correlations have been shown in fish and mammals, as well as invertebrates.

General aspects of this theme were reviewed by Parsons (1990), who emphasized the difficulty of detecting environmental effects under field conditions, because relatively severe stress was needed to detect FA changes even in the laboratory. An initial approach could be to seek effects in ecologically marginal conditions, if these can indeed be determined, so that comparisons of organisms from ecologically benign and presumed stressful regimes could be made. Parsons cited the important study of the satyrine butterfly *Coenonympha tullia* in the Rocky Mountains by Soulé and Baker (1968), in which FA was found to be highest at the highest altitudes where the environmental influences were presumably most extreme. This correlation was based on six characters estimated for adult butterflies from six localities, over the altitude range of 2591–3048 m.

Clarke (1992, 1993) advocated the value of FA for detecting stress in natural environments as it is simple to assess, inexpensive, and widely applicable to a range of biomonitoring and conservation contexts. He discussed two examples to indicate the kinds of application available (see also p. 50).

1. Bushfly responses to avermectin. 'Avermectins' are a group of chemicals which are used widely to control parasites of domestic stock, such as cattle, pigs, sheep, and horses. They can be administered orally or by injection, and recent developments include use of sustained-release devices which liberate the drug into the alimentary canal of the host (Strong 1992). Significant amounts of active chemicals and residues are released in the animals' faeces, and can have severe adverse effects on dung-breeding insects. The drugs do not degrade rapidly from faeces, and breakdown of dungpads by insects may be retarded considerably. Measurements of FA in wing vein lengths of the Australian bushfly (*Musca vetustissima*) reared from avermectin-affected dung revealed higher FA levels than control flies. Additionally, flies collected 4 weeks after treatment of the donor host animals were

significantly more asymmetrical than either 8 week or 11 week samples. By 8 weeks, the avermectin residues did not affect bushfly survival, but a developmental effect clearly persisted beyond this period (Clarke and Ridsdill-Smith 1990).

2. Inbreeding in marine copepods. Clarke's second example involved the rather different context of determining whether the effects of reduced fitness or inbreeding were manifest in FA so that the species might serve as an 'auto-indicator' rather than responding to more clearly external factors. The ramifications of this context are discussed in Chapter 7. The marine copepod, *Tisbe holothuriae*, from Venice was used to explore this premise. In the population utilized, inbreeding constituted a developmental stress in which fecundity decreased, female sterility levels rose, and survival lessened as inbreeding levels increased.

Five meristic characters, counts of spines on the body, were made on both sides of the copepods, and a trend of increasing FA with inbreeding was clear, with four of these five characters significantly different after only one inbreeding generation.

Application of morphological changes such as extent of FA is not yet generally common for biomonitoring, and it is usual to attempt to assess the significance of numerical change, either as relative abundance or presence/absence states, rather than morphological or developmental influences. There has been much discussion over selection of optimal indicator groups of invertebrates in different habitats, with emphasis (with the realization that there is no single 'invertebrate canary' (Hart and Fuller 1974) which can be used to monitor all important disturbances and threats) on seeking a suite of different taxonomic groups whose responses will complement each other to provide more comprehensive monitoring than possible from any one group. Particular invertebrate taxa appear to have significant value as indicator groups, and will conventionally include ecologically sensitive invertebrates occupying different trophic roles in a community. It may, thus, be feasible to develop relatively general methodologies for

Table 2.3 Groups of invertebrates rejected for use as monitors of pollution, recommended by an International Workshop on Monitoring Environmental Materials and Specimen Banking, Berlin 1978 (Luepke 1979)

Group	Rationale
Tardigrades	Biomass small; basic information base inadequate
Nematodes	Difficult to collect; what, if anything, do they indicate? Available information mainly on economically important species
Phytophagous insects	Offer little or no information that could not be obtained from plants
Hymenopterous parasites	Sampling and collection more difficult than for predators; not much additional information, except on specific host–parasite interactions compared to that from more generalist predators
Centipedes	Occupy same functional components as carabids and spiders, but regulate detrital populations rather than pests; not as much known as about carabids and, possibly, spiders

routine assessment of these groups as monitors of environmental health. Conversely, some groups have been suggested as of little practical value for monitoring pollution, on a variety of grounds (Table 2.3). Criteria for selecting indicators may incorporate consideration of many biological levels (Table 2.4).

Indicator groups
One difficulty with nominating indicator groups of reasonably universal (or, at least, meaningful regional) application is the ecological variation which is inevitably present in any reasonably large or widespread group of animals. Some species may show marked and predictable responses to environmental change, but others may not be as sensitive. For *groups* of potential value for indicating community condition, several features are advantageous (Hellawell 1986):

1. The group should be reasonably diverse, but not so overwhelmingly so as to preclude adequate investigation and interpretation.

2. The taxonomy should be well known, with most species described or recognizable. Ideally, keys and field guides usable by non-specialists should be available, or be capable of being produced reasonably easily.

3. The group must be amenable to standard or otherwise defined ecological sampling methods. For short-term active collecting, this might entail being associated clearly with particular habitats, susceptibility to baits or chemical attractants, or simply being active in the daytime, for example.

4. The taxa must be sufficiently abundant that their detection, and detection of changes in incidence and abundance, is straightforward. This condition precludes the use of certain rare groups which might be employed otherwise: see comment on Onychophora, p. 24.

5. The group must be widespread in, and perhaps characteristic of, the ecosystem it is desired to monitor.

6. The main ecological roles of the taxa must be understood, and their responses to the environmental changes it is desired to monitor should be clear-cut. It is likely that a good indicator group will include a high proportion of specialist species.

This character suite contains elements in common with others that have been suggested. By comparison, Jenkins (1971) noted the following criteria for selecting biological indicators, they should:

(1) be cosmopolitan;

(2) abundant;

(3) sensitive to pollution;

(4) show a well-defined response, either
   (a) die or decrease,

Table 2.4 Specific criteria for selection of invertebrate animals for monitoring environmental materials and specimen banking (after Luepke 1979)

1. System characteristics
    i   representative of critical components, functions and processes
    ii   degree of data extrapolation: laboratory to field, season to season,
             comparison between sites, etc.
2. Population characteristics
    i   abundance and expendability
    ii   wide disturbance
    iii population trends
    iv   trophic level
    v   migration pattern
    vi   life table functions
    vii population stability monitoring
3. Individual characteristics
    i   ease of identification (well-defined taxonomy)
    ii   ability to age, class, and sex
    iii body size
    iv   longevity
    v   diet and habitat
    vi   accumulation of compounds of interest
    vii sensitivity to biological effects
    viii exposure to pollutants
    ix   annual cycle characteristics—stages of the life cycle
    x   genetics (active-homozygous for exposure)
    xi   intermediary metabolism allows adequate monitoring
4. Logistical–organizational–other
    i   costs (time, funds, personnel) and maintenance
    ii   collection, preparation, packaging, and transportation in
             consistent standardized manner
    iii scale and resolution of sampling programme
5. Adequate abiotic and biotic information base
    i   land use, air and water quality data, and climatological data
             should be available relative to the species
    ii   baseline data on physiology, reproduction, life table information,
             susceptibility to disease, parasite burdens, and pollutant
             concentrations; must be able to integrate this information with
             other monitoring systems; analysis and evaluation of the combined
             material desirable

(b) change or mutate, or

(c) replace or be replaced by other species;

(5) be non-target species if being used to monitor pesticides; and

(6) have changes that are visible by remote sensing.

The allocation of taxa to 'functional groups' links them by their behaviour or ecological roles, rather than by taxonomic relationships, and can facilitate comparisons between communities with few species in common. As Andersen (1990) noted, the concept is particularly valuable for evaluating ecological systems, and can draw on the knowledge of some of the best indicator groups, such as ants in terrestrial systems. For ants, this approach was pioneered for use in the Australian arid zone (Greenslade 1978) but is now more widespread. Andersen (1990) noted the advantages of this approach, using ants, as:

(1) bypassing our ignorance of biology of individual species because, in most cases, it operates at the generic level;

(2) simplifying complex ant communities by

Table 2.5 Functional groups of Australian ants, relevant to their use in environmental evaluation (after Greenslade 1978)

	Group	Example(s) and relevant features
1.	Dominant dolichoderine	*Iridomyrmex*; abundant, active, aggressive; able to monopolize resources
2.	Associated subordinate Camponotinae	*Camponotus, Polyrhachis*; co-occurring with *Iridomyrmex* and competitively subordinate to it
3. (a)	Hot climate specialists	*Melophorus, Meranoplus*; behavioural/morphological specializations which enable coexistence with *Iridomyrmex*
(b)	Cold climate specialists	*Prolasius, Notoncus*; restricted to cool, wet regions, where influence of *Iridomyrmex* reduced
4. (a)	Cryptic species	Many small ponerines and myrmicines; Forage exclusively in soil and litter; eyes reduced or absent
(b)	Sub-cryptic species	Many small formicines and dolichoderines; forage mostly in soil and litter; eyes larger
5.	Opportunists	*Rhytidoponera, Paratrechina*; extremely unspecialized behaviour; poor competitors
6.	Generalized myrmicines	*Monomorium, Pheidole, Crematogaster*; unspecialized behaviour, but successful competitors because of rapid recruitment and effective defences
7.	Large solitary foragers and/or specialist predators	*Myrmecia, Leptogenys*; unlikely to interact much with other ants because of large size, low population density, and/or specialized diet

condensing large numbers of species into a small number of groups;

(3) providing valuable insights into the major processes operating in ant communities and their associated habitats; and

(4) enabling meaningful comparison between ant communities throughout Australia.

The main classification used involves seven functional groups (Table 2.5) and these can be amalgamated further to give three major components: dominant species and associates (Groups 1, 2), highly interactive taxa (Groups 5, 6), and other species (Groups 3, 4, 7).

Selecting indicator groups and species
Many groups fit some of the desired features of an 'ideal indicator' and have some consequent use in particular habitats or ecological contexts. Thus, for example, freshwater gastropods fulfil two of the five desirable features noted by Harman (1974), namely: (1) they have a relatively long life span; and (2) should be comparatively sessile or not easily able to avoid

stress by migration. Most of them do not fulfil the others: (3) that they should be identifiable easily by non-specialists (a recurring theme in designating and seeking good indicators); (4) that they should be abundant over a large geographic range; and (5) that they should indicate the same conditions or changes over that range. Many gastropod taxa may indicate only the general water characteristics of their normal habitats, rather than the limits of tolerance to various water parameters and changes in those parameters. Most invertebrate groups used and recommended as indicators have their merits and disadvantages in any given context.

'Indicators' can be used in several ways:

(1) most commonly, when a given species is an indicator of a given set of environmental conditions, be they normal or disturbed;

(2) an absence, for example of a stenotopic taxon in a sample may denote a particular disturbance; or

(3) in derivation of an index of community

structure, where 'high richness' or 'high diversity' may indicate little or no disturbance, and the converse.

The twin contexts of 'instant' assessment by analysis of samples taken on one occasion, and continued sampling at intervals over an extended period to reveal change are very different. The former is clearly open to misinterpretation because it may not always be clear if the community revealed is normal or 'stressed'. Indeed, all the above contexts have shortcomings as general models, and interpretation can only be as sound as the level of taxonomic and ecological understanding of the organisms involved will permit. Considering Harman's (1974) appraisal of freshwater gastropods again, for example, in the last of the above contexts:

1. Absence of 'clean-water' species may be a better indication than presence of tolerant ones in estimating pollution as an ideal case. However,
    (a) because biological knowledge of the species involved may be generally poor, reasons for absence may not be clear and the species might well occur in polluted environments elsewhere;
    (b) many clean-water species have very restricted distributions, perhaps only in a few small areas, or a single catchment;
    (c) estimating absence is difficult (p. 77); and
    (d) the investigator must be sufficiently familiar with the local fauna to recognize what is missing.

2. Other factors besides pollution or disturbance can result in low diversity in communities. Lack of understanding of any of these could reduce markedly the applicability of a group to specific or more general problems.

The main value of good indicator invertebrate (or other) species is to reveal the earliest changes or stress to a natural system. Several European mayflies (*Baetis* spp.), for example, disappear from acid-polluted waters before fish living there show any obvious response to the pollution

(Raddum and Fjellhcim 1984), and this sensitivity seems characteristic of many invertebrate animals. Indicator species (as opposed to whole groups) show trends which are not reflected at that time in the whole fauna. In a study of invertebrate assemblages in upland streams in Wales (Weatherley and Ormerod 1990), three streams were 'limed' with resultant increases in pH from 5–5.2 to 6.4–6.9, and increase in calcium and aluminium. A few insect species responded to this shift in stream chemistry but most taxa did not, and the fauna as a whole was thus rather inert. Other examples of invertebrates responding to aquatic pollution are noted in Chapter 3 (p. 50).

The theme of gaining greater ecological understanding by using different indicator groups has been pursued more effectively for terrestrial communities where, for example, several combinations of different arthropod (especially insect) groups have been suggested (New 1984, 1987). A consideration jointly of Collembola (soil and litter, mainly in decomposer food webs), leaf hoppers (Homoptera) and leaf-beetles (Chrysomelidae) (both herbivores, often markedly host specific, but with different feeding modes and habits), ants (a particularly valuable group of insects, because they are virtually ubiquitous and participate in many different trophic interactions (Andersen 1990)) and ground beetles (Carabidae) (active predators), for example, could be used for comparative assessment of many communities and have wide geographical application, in indicating patterns of biodiversity in a widespread group.

The use of Carabidae as indicator taxa, especially in open areas such as fields or grasslands, has been explored extensively in Europe (Stork 1990). The ecology of many species is known in detail, and differences between carabid faunas of different sites can be detected by the single, simple technique of pitfall trapping (Luff 1987). They can thus be used both to detect gross differences between sites and habitats and as an adjunct to classifying and ordinating vegetation associations. Thus, Eyre and Luff (1989) assembled carabid data from 638 heathland and grassland sites in northern and central Europe, which analysis reduced to 17 habitat groups, some geographically restricted and others widespread. Soil water and site altitude appeared to be important factors influencing the distribution of European carabid assemblages.

Management effects on grassland carabids are also evident, with some species favouring intensively managed sites, and others being very sensitive to any changes to their normal natural regimes (Eyre and Rushton 1985; Eyre et al. 1986a,b; Rushton et al. 1989). Carabids have been used in two distinct contexts in relation to responses to pollution (Freitag 1979): the effects of industrial pollutants on the environment, and the application of pesticides—mainly to phytophagous insects of which carabids are predators. Freitag assessed the future role of carabids as pollution indicators as 'boundless', given their biological features and the wide range of chemical parameters to which they variously respond.

The tiger beetles (Cicindelidae or Carabidae: Cicindelinae) have been appraised as a particularly useful indicator group of predatory insects, some of them highly susceptible to habitat change. As well as reflecting regional biodiversity, some may be particularly appropriate indicators of human impact. One additional value of tiger beetles stressed by Pearson and Cassola (1992) is that they can be sampled rapidly and reliably. At a range of sites, in several parts of the world, the first 50 hours of collecting yielded 78–93 per cent of the tiger beetle fauna found in the area. They are also suitable for use by less-experienced workers; in conservation workshops organized by Pearson and Cassola, 80 per cent of students were able to find at least 90 per cent of the tiger beetle species on a study site after only 4 hours of training.

Use of plant-feeding groups as indicators is subject to the caveat that changes in their status may merely reflect changes in vegetation composition, which would be detected more easily by studying the plants themselves. New Zealand tussock montane grasslands are subject to invasion by Agrostis (browntop grass), and this exotic species is associated with the decline of many endemic herb species, and changes in abundance of insect consumers, such as many moth species (White 1987, 1991). The more abundant moth species were, in general, those

that were affected initially by changes in vegetation composition, and White (1987) advocated monitoring of common species as valid in reflecting 'early warning' of possible changes in abundance of scarce species. For management in situations where the requirements of many individual species are unknown, the broader goal of retaining systems, rather than scarce species, may benefit all (or most) of the species present (p. 82), by analogy with the 'umbrella' concept.

In Austria (and elsewhere), the deterioration of alpine habitats through intensifying winter sports activity is a matter for general concern. The hoverfly (Diptera, Syrphidae) fauna of a natural meadow and an intensively-skied meadow on Stubnerkogel were compared by Haslett (1988). With sufficient knowledge of the individual species it was clear that greater numbers of generalist and grass-feeding species were present on the ski-slope, whereas more specialized foragers were few (Fig. 2.4). Such differences (even when only relatively small numbers of species are involved: in this instance 15 species on the ski-slope, 17 on the 'control' slope) are ecologically informative and, again, such changes—even though the richness and diversity of the two sites may be very similar— may be an effective early warning of significant community change.

Habitat change, though, inevitably engenders faunal change, and many such changes may not necessarily be of conservation significance. Using a range of different potential indicator taxa, rather than relying on one or few species in any given context (which can lead to an overly narrow focus), produces a finer level of information (Noss 1990). The diversity of invertebrates allows for the distribution of many species, and their environmental relationships, to be explored simultaneously, using ordination techniques (such as those discussed by Gauch 1982). These may permit the response of a group of taxa to be explored for use in environmental monitoring, as Kremen (1992) has undertaken for butterflies in Madagascan rainforests. The protocol she defined is applicable widely to assessing invertebrates as indicators, and in determining appropriate indicator groups objectively rather than according to the (often) more subjective predelictions of particular workers, or merely by following precedents of unknown or doubtful value. Kremen evaluated the indicator properties of butterfly assemblages against an environmental pattern of disturbance and montane/topographic gradients (Table 2.6). Details of analytical methods are given in her paper; basically, canonical correspondence analysis was used to examine relationships between environmental parameters and species distributions; detrended correspondence analysis was used to interpret species abundance matrices of butterfly and plant data for the transect samples (butterflies) or plot samples (plants). Linear regression was used to seek relationships between local butterfly diversity and plant species richness or diversity.

Kremen's main findings (Kremen 1992) were that the butterfly assemblages were:

(1) excellent indicators of heterogeneity arising from the topographic/montane gradient;

(2) limited indicators of heterogeneity due to disturbance; and

(3) poor indicators of plant diversity

The various approaches to using invertebrates in assessment of environmental quality were addressed by Eyre *et al.* (1986), who emphasized that knowledge of some taxonomic groups was indeed sufficient for this. Ground beetles (Carabidae) and spiders from a suite of different terrestrial habitats, and water beetles from aquatic habitats, in Britain and Europe

are examples. The two ordination techniques employed, TWINSPAN, or two-way indicator species analysis (which divides the data progressively into two groups, to provide a hierarchical classification which can be interpreted ecologically if indicator characteristics of the species are understood) and DECORANA, or detrended correspondence analysis (which orders the species combinations present in the data set, and interpreted by environmental data from the sampling sites), have become commonplace for categorizing community groups in diverse surveys. It is necessary to identify the taxa to species level (or, at least, to recognize them consistently to 'morphospecies') for these methods to be applied, and the ordinations obtained can be interpreted properly only from sound knowledge of the ecology of the invertebrate group concerned. However, well-known invertebrates are of considerable value because of the numbers of species and individuals available.

The converse use to invertebrates as indicators of 'damage' is that particular groups may respond rapidly to ecosystem rehabilitation and restoration, a facet of increasing importance to (for example) mining and forestry practices in many parts of the world. Clear trends of faunal change may occur with continued sampling, with progressive return of native species. On North Stradbroke Island, Australia, Majer (1985) showed a linear pattern of ant species return to rehabilitated sand mines over 7 years, for example. Majer had earlier (1983) (Table 2.7) suggested that the potential applications of

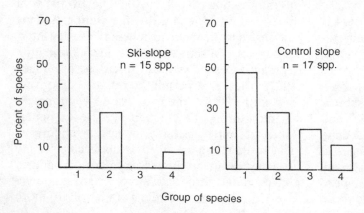

Fig. 2.4 Change in assemblage composition of hoverflies (Diptera, Syrphidae) from a 'natural' slope to a modified ski-slope in Australia, emphasizing how (although overall diversity may change little), the ecological characters of the assemblage may change substantially. 1, Generalist foragers; 2, grass-foragers; 3, specialist species; 4, unknown habits. (After Haslett 1988.)

Table 2.6 Environmental variables used by Kremen (1992) for canonical correspondence analysis (CCA) of butterfly data in Madagascan rainforests

Environmental variable	Nominal variable	Description
Topography	0	Ridges
	1	Slopes
	2	Streams
Disturbance	0	Selectively logged 25 years ago
	1	Selectively logged 2 years ago
	2	Slashed for logging trails
Altitude		m, measured
Plant species richness		Margalef's index
Canopy height		m, estimated
Floral abundance		Number of individuals in flower (averaged over time)
Floral richness		Number of species in flower (averaged over time)
Butterfly dominance		Berger–Parker index[a]

[a] $d = \dfrac{N\text{max}}{N}$, where Nmax is the number of individuals of the most abundant species.

ants as indicators were, indeed, very widespread for assessing areas to be disturbed and how these are being restored:

(1) to provide baseline data needed to determine goals for eventual restoration;

(2) to monitor the extent of ecosystem recovery in rehabilitation;

(3) to monitor ecosystem degradation in areas which are subject to disturbance;

(4) to interpret the conservation status of nature reserves;

(5) to understand the faunal affinities between different reserves or sites.

Various similarity or comparison indices are applicable in interpreting such data, and optimal suitability of these differs according to the context. Need for rehabilitation planning after commercial or other ecological disturbance to a site is now acknowledged (and demanded) widely, and some examples are given in Chapter 4. In general, the aim is to restore the system to the condition approximating that before it was disturbed. There is thus a practical need for sound baseline information to be obtained

before the disturbance, and control sites to be established for detecting future trends; this itself has frequently not been undertaken in the past. The subsequent need is to determine groups of organisms (such as ants, as above) whose features can monitor progress of a rehabilitation process towards a natural assemblage, with the provision for the process to be altered if necessary to enhance ecological integrity.

Development and refining of use of invertebrates as indicators is one of the most important aspects of community monitoring, and will continue to be applied in many contexts in conservation management and assessment. This section

Table 2.7 Attributes, suggested by Majer (1983), which make ants (Hymenoptera: Formicidae) good biological indicators

They are extremely abundant
There is high species richness
There are many specialist species
They occupy high trophic levels
They are easily sampled
They are usually easily identified
They are responsive to changing environmental conditions

has demonstrated some important aspects of this, in relation to a number of external threats to the invertebrates and their broader biological context. These threats are enumerated more fully in the next chapter.

FURTHER READING

Collins, N. M. and Thomas, J. A. (ed.) (1991). *The conservation of insects and their habitats*. Academic Press, London.

Hawksworth, D. L. (ed.) (1991). *The biodiversity of microorganisms and invertebrates: its role in sustainable agriculture*. CAB International, Wallingford.

Mattson, W. J. (1977). *The role of arthropods in forest ecosystems*. Springer-Verlag, New York.

New, T. R. (1984). *Insect conservation: an Australian perspective*. W. Junk, Dordrecht.

Norton, B. G. (ed.) (1986). *The preservation of species*. Princeton University Press, Princeton, NJ.

Rosenberg, D. M. and Resh, V. H. (ed.) (1993). *Freshwater biomonitoring and benthic macroinvertebrates*. Chapman & Hall, New York.

Samways, M. J. (1994). *Insect conservation biology*. Chapman & Hall, London.

Stork, N. E. (ed.) (1990). *The role of ground beetles in ecological and environmental studies*. Intercept, Andover.

3 A CHANGING WORLD: THE THREATS TO INVERTEBRATES

INTRODUCTION

'Human interference, be it habitat alteration to complete destruction; chance and deliberate introduction of predators or competitors; or exploitation by collecting activities, is the summary reason for the crisis of twentieth-century extinctions in all groups of organisms' (Solem 1990).

Solem encapsulated succinctly much of the need for conservation in this statement about the Hawaiian land snails and went on to comment that much of the damage that has occurred can no longer be remedied. Much of the practice of conservation is indeed concerned with the detection and 'rescue' of threatened taxa, communities and habitats, but also with preventing others from reaching that parlous state and attempting to predict the effects of human activities and, by some means, regulating these or controlling them in ways that can prevent further serious losses of biota. The elucidation of 'threatening' processes and events is vital, and abating their effects is a major facet of practical conservation.

The diversity of processes threatening to invertebrates is immense, with virtually any nuance of habitat or community alteration potentially affecting some sensitive or specialized resident species.

Changes in ecosystem condition can be represented as in Fig. 3.1. This scheme (IUCN, UNEP, WWF 1991) emphasizes the accelerating loss of natural capability to sustain these complex systems as they are changed progressively by human activity. Natural ecosystems provide all the support needed for their resident biota and are potentially fully sustainable, whereas this capability is reduced with change. 'Modified systems' are defined as those where human impact is greater than that of any other species but whose structural components are not cultivated; examples are naturally regenerating forest used for timber production. Cultivated and built systems manifest greater impact. All have the potential to become 'degraded', where diversity, productivity, and habitability have been reduced, sometimes massively, and where rehabilitation is needed to render them sustainable.

CATEGORIZING THREATS

The major threatening processes to invertebrates can be grouped into four very broad categories (Wells *et al*. 1983), each of which can be subdivided extensively:

1. Effects of habitat destruction or change.

2. Effects of pollution and pesticides.

3. Effects of exotic species.

4. Effects of overexploitation and overcollecting.

These are discussed separately below.

The range of threats to any given invertebrate group or community is likely to include an amalgam of these, although one or other usually appears to be of paramount importance. Some appreciation of the variety of threats which can be defined broadly for any given taxon is

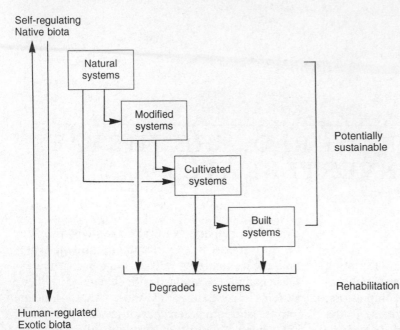

Self-regulating
Native biota

Human-regulated
Exotic biota

Fig. 3.1 General scheme of habitat change by people and broad consequences for biotic change and need for active management to maintain these (IUCN, UNEP, WWF 1991).

exemplified by a simple listing of the human-induced changes that influence survival of the British dragonflies (Chelmick *et al.* 1980):

(1) loss of water bodies;

(2) modification of ditches and rivers—such as influencing water levels, scooping out vegetation, creating sharp margins, or altering slopes of banks;

(3) drainage;

(4) making fluctuating water levels by pumped drainage schemes and reservoirs;

(5) use of chemicals against mosquitoes, and wind drift of terrestrial insecticides;

(6) use of herbicides in weed clearance, including those applied against margin vegetation;

(7) pollution, including run-off of agricultural fertilizers

(8) overmanagement by amenity or fishing interests;

(9) overstocking with fish or ducks, which may feed on aquatic larvae; excessive droppings from birds can lead to eutrophication of water bodies;

(10) loss of immediate natural surroundings, including shelter belts;

(11) loss of modification of wider hinterland;

(12) lack of management, resulting in shading or choking of water with silt and plants; and

(13) afforestation leading to excessive shade and lowering of water bodies.

Dragonflies (Odonata) depend on the well-being of aquatic systems as larval habitats, and the nature of the surrounding area as adults, since these may move over at least several kilometres; as implied above, changes to either the freshwater systems themselves or to surrounding terrestrial areas may contribute to their decline. The above listing includes a wide variety of impacts, from absolute loss of habitat to the rather subtle and less obvious effects of changes to nearby areas important as a buffer to protect the primary breeding areas.

In this case, and many others, features of nearby habitats, which do not appear initially to be relevant to invertebrates, may be very important. They can be assessed only from good ecological knowledge. The role of vegetation for

Table 3.1 Possible roles of vegetation for dragonflies (after Buchwald 1992)

Single plants or plant stands
  1. Larval habitat: feeding, protection from predators
  2. Oviposition site or substrate
  3. Substrate for hatching
  4. Perch site: territorial behaviour, hunting, waiting for mates, mating, sun-basking, roosting
  5. Boundary demarcation of territory
  6. Protection from predators; shelter during inclement weather
  7. Signal releasing habitat selection (specific plant species)

Pattern of plant communities/surrounding landscape
  8. Feeding area
  9. Maturation area for immature adults
10. Influences on microclimate
11. Signal releasing habitat selection ('aspect' of vegetation in surrounding areas)

dragonflies, can be very complex, and the 11 distinct functions of vegetation in relation to the ecology of adult dragonflies noted by Buchwald (1992) (Table 3.1) indicate the subtle influences that changes may have on the population's well-being. Maintenance of this (or an equivalent) range of natural functions may necessitate management of such adjacent habitats in addition to the primary one. Lack of information may permit inadvertent harm to occur, and the key to successful management (p. 89) is indeed sound biological knowledge of the taxa and systems involved.

Apparently unrelated factors, such as conifer plantations adjacent to water bodies, can induce marked changes in stream ecosystem function, and influence the suitability of the habitat for many aquatic invertebrates (Weatherley *et al*. 1993). Suites of threatening factors could be nominated for virtually any invertebrate group, and the above example simply emphasizes that gaining even a reasonably complete picture of threats can be immensely complex, especially for taxa whose ecology is only poorly known. Different factors may differ locally in priority, but embracing lists of the kind given here for dragonflies provide a useful check of 'what to look for'. They exemplify some of the major threat categories that can occur in many situations:

1. Physical destruction or alienation of the habitat: proximal effects.

2. Biotic contamination through introduction of alien species, or by manipulating the natural balance of species already there: proximal effects.

3. Alienation or destruction of peripheral or nearby habitats with unknown relevance to the main target habitat: intermediate effects.

4. Pollution effects, which may be adjacent to, or far distant from, the sites they eventually affect.

Much of the practical consideration of invertebrate conservation (Chapter 4) involves detecting and removing such threats, or protecting particular taxa or sites from them and from possible overexploitation. It is often difficult to predict the consequences of possible changes, because the intensity of a particular environmental disturbance may be critical, and the threshold intensities at which major damage may occur are largely unknown. Part of countering any threat is being able to assess the likely consequences of *not* countering it, and the degree to which it should be countered or regulated. Indeed, even detecting a potential threat can be problematical, and many apparently innocuous local disturbances may prove to have serious and far-reaching effects as their ramifications become more obvious. Chemical spillage or waste discard, for example, can influence invertebrates many kilometres downstream from where a small amount of pollutant is released into flowing water.

The following discussion shows many examples

able 3.2 Factors which may contribute to a species' scarcity and susceptibility to extinction (after Ehrenfeld 1970), and their relevance to invertebrates

Factor	Comment
Specialized habitat preferences	Widely applicable
Restricted distribution	Widely applicable
Intolerance of presence of humans	Manifest in habitat change and other anthropogenic effects rather than presence *per se*
Species reproduction in few large aggregates	Scarcely applicable (? social insects)
Low litter size, slow maturation, long gestation period	Fecundity often high, mainly fast maturation
Large size and predatory habits	Predators vulnerable, especially if monophagous or oligophagous
Behaviour patterns less adaptable to changing conditions	Many sedentary or with low vagility; many 'specialist' taxa
Excessively hunted or trapped	Low proportion of taxa

of threats to particular invertebrates. However, it is impossible to categorize simply all possible threats implied by habitat loss or other changes to the local environment: in essence, they encompass much of human activity, especially if carried to 'excess'—itself difficult to define except in terms of the undesired results visible only with hindsight! Consider, for example, some of the possible changes to the fauna of an area of estuary or sea into which heated effluents are discharged (Naylor 1965). Biological consequences could include:

(1) elimination of some species, especially those adapted to living in cool water;

(2) increased abundance of other species;

(3) facilitated colonization or invasion by warm-water species;

(4) increased abundance of fouling or boring species; and

(5) facilitation of establishment of warm-water species in temperate regions.

In addition to these changes to the community, dispersal of heat may result in changed salinity gradients and flushing patterns—thus affecting the dynamics of estuarine systems and, possibly, the behaviour of migratory species, such as crustaceans.

In general, many different facets of a species' biology and habitat needs may render it susceptible to extinction (Table 3.2).

HABITAT DESTRUCTION OR ALIENATION

Without a suitable place to live, an organism (or species) cannot survive. Many different scales are involved in defining suitable habitats. Within a large area, a given species may only be able to thrive in a mosaic of small patches with particular topography, microclimate, or biotic associations. Particular landscape elements may, therefore, be the critical habitat needs for a species and, as Samways (1994a) has stressed, the heterogeneity characteristic of natural systems is a major component of habitat structure. Accelerating and large-scale change of the natural ecosystems of the world to satisfy the needs of burgeoning human populations ensures that, as demands for food, land, mineral resources, and trade goods continues to grow, no part of the world remains totally free of resultant intrusion and exploitation, or of the consequences of this. Habitat loss is therefore the paramount threat to many animals and plants, and the speed and intensity with which this can occur leaves little chance for many species to adapt gradually to the changed conditions. This section focuses on direct physical habitat change by human activity and its consequences: the other threats also have effects on habitat but, for example, 'pollution' is regarded generally as the undesirable by-product of direct intrusion rather than habitat alienation *per se*.

Habitat alienation takes many forms and, as with other threats, may not seem to be large at first. It contains the twin elements of habitat loss, and insularity, or increasing fragmentation of habitat patches. Both elements contribute to decline in biological diversity (Wilcox and Murphy 1985).

It follows that many threats can be very localized, and many species, assemblages, and habitats of high conservation value are, of course, also highly localized. Although not of conservation significance in themselves, the following examples of apparently minor intrusion by people seeking outdoor recreation indicate the kinds of local events that could, on occasion, constitute threats to rare species:

1. Trampling of moorland peat can kill large numbers of larvae of the moorland cranefly (*Molophilus ater*, Tipulidae) in Britain, and ground rendered bare by people walking on it might be less attractive to the flies for oviposition (Bayfield 1979).

2. Invertebrates of grassland litter appeared to be affected adversely by much lower levels of trampling than are needed to produce floristic change (Duffey 1975). Studies of fauna of litter (enclosed in bags) differed considerably between controls (untrodden litter bags) and bags subjected to five treads or 10 treads each month for a year. Declines in abundance with these regimes occurred in Mollusca, Isopoda, Araneae, and several insect orders, but differences between the two levels of treatment were usually small. Dipteran larvae *increased* in abundance, but most other groups were reduced to around half their normal (control) abundance (Duffey 1975). Seventeen of the 21 most abundant beetle species declined in this way, two others changed little, and the other two became more abundant. There was thus considerable species-level variation in sensitivity to treading.

Visitor trampling on coastal rock platforms or coral reefs (Liddle 1991), without further disturbance, are other facets of apparently 'simple' recreational activities which may cause damage. Another is the skiing, alluded to on p. 35, where floristic change might be induced and this, in turn, affects the spectrum of invertebrate consumers able to live in sensitive alpine areas. In coastal New South Wales (Australia), recreational activities affecting local marine coastal organisms include fishing, snorkelling, diving, collecting for food or aquaria (p. 13), or simply walking around (Underwood and Kennelly 1990), and these activities collectively have large but currently unknown potential to disrupt intertidal communities.

Such seemingly harmless activities are taken for granted by many conservation-minded people as a normal facet of enjoying outdoor environments in an environmentally aware way, but the examples are salutary in indicating the extreme sensitivity of many invertebrates to disturbance of their habitat. Management (p. 89) may necessitate regulation of traffic density, or human activity to be restricted to particular areas in some sensitive habitats.

Even this can prove problematical. In many National Parks, human intrusion is indeed regulated by restricting vehicles to a limited range of roads or tracks. These may also be used preferentially by animals: in the Addo Elephant National Park, South Africa, elephants now use roads much more than previously; they defaecate to an increasing extent on road surfaces, rather than in the bush, and visitors' vehicles now run over increasing numbers of an elephant dungbeetle, *Circellium bacchus* (Scholtz and Chown 1994). The beetle is flightless, and an important species in the decomposer/recycling system, and there is now concern for its conservation status in the park. Decline has been prompted also by a reduction in numbers of the main alternative 'host', buffalo, with consequent lack of food for the beetles.

Quite clearly, control of all such habitat changes is undesirable and impossible, would alienate much public goodwill, and would distract attention from much more damaging and far-reaching changes involving massive levels of habitat despoliation. Many small changes such as those noted above are, indeed, unimportant because they do not affect the existence of species or populations, but they demonstrate the care and subtlety that may be needed in the management of particular small habitats or sites critical for particular notable species (p. 75). They become significant in the conservation of small or highly localized invertebrate species. However, some general points have emerged regarding the conservation of invertebrates that depend on particular resources in visitor-prone areas. Oversanitation of woodlands by removal of fallen or waste timber, for example (perhaps for firewood), may affect the large number of species that need dead wood to survive (Speight 1989), and such microhabitats can easily be neglected in conservation management unless their roles can be emphasized sufficiently.

Most concern over habitat destruction is for larger-scale events and for more persistent gradual declines in the extent of natural areas constituting the familiar avenue to extinction through habitat loss, perhaps initially at local levels, but then to a regional or global scale. Very generally, terrestrial invertebrates (as other biota) may be affected by human activities associated with deforestation, agricultural conversion or intensification, surface mining, industrialization, urbanization, and any other process that results in (or leads to) changes in natural vegetation, topography, or soil condition. Freshwater taxa may lose habitat through impoundment, drainage, channelization, and factors allied to nearby agriculture or forestry, which increase siltation or exposure. Local 'sanitation', such as clearing waterside vegetation, may also be important, as noted earlier for woodlands. Marine invertebrates are subject to habitat loss along coasts (areas which typically have high adjacent human populations) and are susceptible to the demands of tourism and recreation, industrialization, dredging, reef damage, removal of coastal vegetation (such as mangroves), and others. Direct habitat change to the open oceans (other than by contamination—pollution, p. 48) is not as widespread, although many coastal effects are highly relevant to those many pelagic invertebrates that pass part of their

life in shallow waters. Collectively, the above and related effects incorporate direct threats to most kinds of invertebrates, with the greatest magnitude of effect perhaps being on non-marine systems.

The physical presence of suspended matter from erosion, for example from nearby forestry, agriculture, and mining activities, can lead to several different potentially harmful effects on freshwater aquatic invertebrates, and, in due course, on marine systems such as coastal reefs. For fine inorganic sediments in New Zealand streams resulting from alluvial gold-mining discharges, five main mechanisms may occur:

(1) reduced light penetration into turbid water, leading to reduced primary production—the basis of aquatic foodwebs;

(2) reduction in quality of the stream epilithon (phototrophic biomass) as food for invertebrates;

(3) clogging or infilling of streambed gravels, destroying habitat for many cryptic benthic animals, and reducing interstitial dissolved oxygen

(4) avoidance reactions—such as increased drift of invertebrates leading to removal of high proportions of residential populations;

(5) accumulation of particles on body surfaces and respiratory structures.

In a study on six streams in the South Island of New Zealand (Davies-Colley et al. 1992; Quinn et al. 1992), benthic invertebrates had significantly lower densities at sites downstream from mine discharge entry, and densities were negatively correlated with water turbidity. In the stream with highest turbidity loads, taxonomic richness was also significantly lower downstream.

Whereas we hear much about 'destruction of the rain forests' or of coral reefs in the tropics, habitats that are vital in harbouring some of the highest-known diversities and densities of invertebrate animals, and which inevitably result in massive loss of species, many smaller-scale events are also very important in emphasizing the paramount value of habitat protection in conservation. Deforestation of even small areas on isolated islands, or of remnant woodland patches on continents, for example, can exterminate endemic or local invertebrates. Wells *et al.* (1983) emphasized that the distribution of some species is so restricted that they could be eliminated by single events such as a single timber concession or building a factory. Many invertebrate species have viable breeding populations that occupy only a hectare or less. Amongst British butterflies 15 species can form closed populations (those with no natural interchange of individuals with other populations) with a minimum breeding area of 0.5–1 hectares, and 11 further species can survive on 1–2 hectares (Thomas 1984).

Two further examples are given to illustrate the restricted nature of the habitats of some terrestrial invertebrate and their subtle resource needs:

The South African acanthodriline earthworms

These comprise around 90 endemic species, with the genus *Udeina* being particularly diverse. They are typically forest taxa and depend on wet, organically rich, cool soils, mainly in moist habitats (Ljüngstrom 1972). They are gradually being exterminated because of destruction of forests, which had been reduced by 97 per cent by the 1950s (Ljüngstrom 1969). Most remaining natural forest stands are small, and soils under exotic forests do not support native earthworms. Deforestation may be followed by burning, which alienates the habitat further, and cultivation then reduces worm abundance even more. Ljüngstrom (1972) believed that acanthodrilines could not colonize even undisturbed native grassland heaths, probably because these have different soil temperature and moisture regimes, and that many of the worms had probably become extinct in the past few centuries.

The no-eyed big-eyed wolf spider, Adelocosa anops

This is an obligate cave-dwelling blind spider which is restricted to a single cave system on Kauai, Hawaii. It forages actively for invertebrate prey, and is known to live in small cavities and crevices of a single lava flow series (Howarth 1979). *Adelocosa* is one of only two obligatory cave lycosids known. Wells *et al.* (1983) cited withdrawal of ground water, and its pollution, as the greatest threat to this desiccation-sensitive spider and other members of the cave invertebrate community. Increasing tourist development in the area, including planning for hotels and roads, has also occurred. The largest lava cave in the locality was ruined by being covered with 5 m of waste sugarcane residue (Howarth 1979), and surface deforestation has eliminated tree roots in the cave.

More intrusive habitat change also occurred. The caves are subject to human visitors, and are Civil Defence shelters. Though the spider might be able to survive in crevices, its 'buffer' areas and the community that it represents are being changed substantially by this complex array of destructive effects (Howarth 1981).

The need to counter the pervasive effects and potential of habitat change, constituting a very high proportion of invertebrate conservation activity, renders selection of examples for discussion somewhat arbitrary. Habitat loss constitutes a universal threat, but is often combined with others, noted later in this chapter. Thus, although many aquatic invertebrates (including some known from single small areas) are indeed threatened by habitat destruction, they are subject also to effects of pollution and human exploitation, and one or other of these may be the more important in many instances. Many examples emerge in this book.

Table 3.3 summarizes the threats to a number of invertebrate groups cited by Wells *et al.* (1983), and attempts to rank their importance—sometimes rather subjectively. Although habitat change emerges as a virtually universal threat, very little information is available on the specific threats to many invertebrate groups. These include free-living animals, such as the marine Pycnogonida (sea spiders), Chaetognatha (arrow worms), Ctenophora (comb jellies), and Priapulida (priapulids) (all of which contain geographically restricted species), the predominantly freshwater Rotifera (wheel animalcules), the ubiquitous Nematoda (roundworms), and the parasitic 'worm' groups in their entirety.

Table 3.3 Threats to invertebrates (summarized from Wells *et al.* 1983).

Taxon	Threat category			
	Habitat change	Pollution	Exotic species	Over exploitation
Protozoa	x	xxx		
Porifera	x	xx		xx
Cnidaria	xx	xx		xx
Platyhelminthes	x	xx		
Nemertea	xx			
Mollusca				
marine	x	xx		xxx
freshwater	xx	xx	x	xx
terrestrial	xx	x	xxx	xx
Annelida	xx	x		x
Arthropoda				
Merostomata	x	x		xx
Arachnida	xxx	x		x
Crustacea	xxx	x	xx	xx
Insecta	xxx	xx	xx	x
'Myriapoda'	xx			
Onychophora	xx			x
Tardigrada	x	x		
Bryozoa	x	xx		x
Brachiopoda	xx	xx		x
Echinodermata	x	xx		xxx

xxx, Predominant or most frequently cited threat; xx, significant or widespread threat; x, local threat, or restricted to few contexts.

It is indeed sobering that such vast sections of the invertebrate spectrum cannot yet be assessed meaningfully to determine even their basic conservation needs, a theme which is developed in Chapter 8.

The above examples, and many other possible ones, demonstrate the importance of habitat at two levels:

(1) the main 'system' occupied by the species under discussion; and

(2) the 'microhabitat' needs, buffered within that system.

The latter may be very precise and determine absolutely the range of the species. Thus the first aquatic insect to be proposed for addition to the *United States list of endangered and threatened wildlife and plants* was the Wilbur Springs shore bug, *Saldula usingeri* (Heteroptera: Saldidae), then known only from one hot spring area in California. In order to thrive there, the bug must cope with high temperatures, high sodium and chloride concentrations, and (more unusually) high lithium concentration (Resh and Sorg 1983). A detailed search revealed *Saldula* populations also in several other springs within and outside the drainage area where it was discovered. Predictions of its distribution on the basis of the above water chemistry parameters were largely fulfilled, indicating that the specialized water characters determined a suitable microhabitat. A related aquatic bug, the Ash Meadows naucorid, *Ambrysus amargosus* (Heteroptera: Naucoridae), occurs in a single set of thermal springs in Nevada, and has been endangered by disruption of the springs through: (1) water removal for irrigation, (2) overgrowth by vegetation and (3) disruption by constructing refugia for an endangered fish (Polhemus 1993).

The need for such precise and specialized

microclimates and small-scale habitat structure augments the value of many invertebrates as subtle indicators of environmental quality (p. 25) and differs from the demands of most larger animals (such as mammals) which, despite specialized needs, may operate predominantly at a larger 'scale'. Assessing the effects of larger-scale changes on the microhabitat needs for invertebrates is generally difficult, and often overlooked by people accustomed to catering for the needs of larger biota. One common context is widespread clearing of tropical forests for agriculture, either for local shifting agriculture or for broader scale commercial development. The effects of associated soil compaction, erosion and similar processes on soil arthropod faunas are usually unclear. Soil invertebrate diversity can be very high, even in temperate regions (Lattin 1993). A preliminary study in Nigeria (Lasebikan 1975) compared the soil arthropods of two sites, one cleared and one uncleared, at intervals of 3, 8, and 10 months after clearing. There was a strong suggestion of faunal reduction over this period, both in number of species and in overall density. Only 7 of 48 families or orders increased in density on the cleared site, whereas 16 declined. The number of groups present approximately halved in the cleared habitat. The effects of that clearing, applicable much more generally, included:

(1) exposure of the soil surface, changing the temperature and moisture regimes;

(2) increasing impact of rain on the surface, affecting soil structure and intensity of leaching;

(3) reduction of litter cover, affecting microclimate, food supplies, and insulation; and

(4) loss of humus, and effects on the microbiological populations of underlying soil.

'Case-histories' for particular invertebrate conservation needs emphasize repeatedly the need to consider such 'fine-grain' effects, and the importance of autecological and synecological studies which can determine these as a basis for effective habitat management for those species.

In general, the more ecologically specialized invertebrates are the most susceptible to localized habitat destruction. A short-term study of the effects of slashing and burning Mexican tropical rainforest on the ant fauna by MacKay *et al*. (1991) showed that, a month after burning, species richness was reduced by more than half, leaving only the relatively generalist taxa thriving in the area.

Once habitats become lost or fragmented (as is occurring most dramatically with tropical forests: Janzen 1986; Sutton and Collins 1991), the future for many invertebrates outside protected areas (p. 82) becomes increasingly bleak. Many ecologists have echoed the sentiments of Frankel (1976) to the effect that few, if any, natural habitats will remain outside protected areas within a few decades. The likelihood is that many such habitats will indeed persist (or, in temperate regions, particularly, may be re-created: p. 96) but will be too small and too isolated to assure the sustainable existence of many species and communities. Invertebrate populations, *inter alia*, may then become 'latently extinct'. Sutton and Collins (1991) noted the example of Bukit Timah (Singapore). This isolated small primary forest area may be subject to loss of many of the tree species because of lack of facility for cross-breeding, so that the present generation of dominant vegetation may, effectively, be the last one for the area and local extinctions are likely. In turn, the extinction of insects and others which depend on those trees—including specific herbivores and pollinators, and specific natural enemies of these—is inevitable also. This process might take another century or so, but is already in train. The major interpretative problem is that *current* threats have failed to cause a marked diminution of species diversity, so such systems may be adjudged undamaged or secure. Janzen (1986) referred eloquently to such taxa as 'living dead'.

POLLUTION

Pollution is the consequence and by-product of habitat change, especially that fostering industrialization, urbanization, or intensive agriculture, and, in a broad sense, refers to the introduction of alien substances into a natural environment or the change in balance of chemicals normally present there, which has a detrimental effect on the organisms living there. It is commonly thought of in terms of pesticides and industrial wastes. Broadly it can refer to physical, chemical, or biological changes, but is most familiarly restricted to chemical effects or input of energy (such as heat) and it has been the subject of numerous texts (such as Hynes 1960; Carson 1963). It can occur in any facet of an environment—acidification of rain engendered by industrial smoke-release, soil contamination by accumulation of chemicals applied to agricultural systems or as 'waste', and deliberate or accidental release of chemicals or complex compounds such as oils into aquatic regimes are only some examples of the broad range of possible causes. Pollution may thus result from deliberate application of chemicals (such as pesticides or fertilizers), accident or inadvertent action, or the deliberate or unheeding discard of wastes. The major effects are of global concern and the long-term climatic changes thought now to be in train (p. 159) are among the more sobering outcomes. Disposal of industrial wastes is likely to remain a formidable problem for humanity.

Traditionally, the oceans have been used as a disposal site for a wide range of hazardous wastes, as well as sewage being released close to coasts, but many biologists fear that the capacity of the marine system to absorb these is already being exceeded. Little specific documentation of the effects of such 'dumping' on oceanic invertebrates is available.

Two kinds of biotic response to pollution are common (McCahon and Pascoe 1990), depending on the intensity and persistence or duration of the intrusive events:

1. A sudden abrupt change, such as an oil spill or single industrial chemical release or pesticide application, can result in rapid (acute) effects to organisms.

2. More gradual effects, such as exposure to low (non-lethal) concentration of pollutants may eventually engender effects, such as serious diseases or malformations, leading to reduced fitness and/or death over a long

period of time, as chronic effects. This category includes many instances of aquatic insect deformity, enabling use of the taxa as indicators (p. 28).

They exemplify the two major kinds of disturbance (perturbation) to natural ecosystems, commonly referred to as 'pulse' and 'press', respectively. The 'pulse' disturbance is one that causes an 'instantaneous' change in a community, commonly through inflicting sudden mortality, followed by a period when the community is free of further similar disturbance and during which it may recover. A 'press' disturbance, by contrast, is a sustained disturbance which is likely to lead to permanent changes in community structure.

The effects of pollution on biotic assemblages may differ fundamentally in terrestrial and aquatic environments (Heliövaara and Väisänen 1993). Pollution in terrestrial systems is commonly a gradual process, such as the build-up of industrial effluents in air or of metals in soil. By contrast, much pollution in aquatic (particularly freshwater) ecosystems is 'sudden', whereby toxins may abruptly change the composition of invertebrate assemblages, or cause large-scale death.

As with habitat alteration, pollution effects are potentially universal. The brief discussion here stresses the susceptibility of invertebrates in a number of general and more specific contexts, and augments the earlier discussion of indicator taxa (p. 27) by enumerating a few more relevant examples.

An oribatid mite, *Humerobates rostrolamellatus*, is a common species in European orchards, and its ecology has been studied extensively. It is very sensitive to sulphur dioxide, and has thus been used as a reliable indicator of air pollution. In this context, its rapid response time (around a week) gives it considerable practical advantage over some other air pollution indicators, such as lichens, whose responses may take several months (André *et al*. 1982).

Soil and litter arthropods

Pollution by toxic metals is widespread in terrestrial ecosystems, but may not be as 'obvious' in rural or non-urban areas as in heavily industrialized zones. Terrestrial isopods (woodlice, or slaters, *Porcellio scaber*) have been found recently to be sensitive indicators of cadmium and lead pollution (Dallinger *et al*. 1992), allowing metal contamination to be detected on a 'fine-grain' scale by analysing concentrations in body tissues. Earlier work on isopods had shown them also to be useful monitors of copper contamination of soil and litter (Wieser *et al*. 1977), and for monitoring the above three metals and zinc (Hopkin *et al*. 1986, 1989, using *P. scaber* and *Oniscus asellus*). These studies, though, were in the higher contaminated zones subject to industrial effects, and Dallinger *et al*. (1992) confirmed their indicator value in less contaminated regions. Significant correlations between lead and cadmium concentrations in *P. scaber* tissues and environmental concentrations were found at all 22 sites sampled around Innsbruck (Austria) and these correlated generally with traffic density, the source of the petrol constituents. Local variations occurred between 'exposed' sites (near roads) and 'non-exposed' sites (separated from roads by vegetation or buildings), and lead concentrations on arterial roads were correlated with proximity to the city and with high traffic density. Dallinger *et al*. (1992) emphasized that in urban environments only a few kinds of animal are amenable to such use as sensitive biological indicators of metals. For isopods, as for Diplopoda (Köhler *et al*. 1992) and some other litter-feeding arthropods, tolerance of heavy metals may vary considerably among species. Feeding efficiency (assimilation) of millipedes decreased after exposure to high lead concentrations (Köhler *et al*. 1992). They are among the most important decomposer groups in the soil macrofauna (Swift *et al*. 1979), about half their diet consisting of the bacteria and fungal hyphae, which accumulate heavy metals strongly, rather than feeding on decaying leaves alone.

Thus, although the overall abundance of an animal group may remain relatively constant in polluted conditions, relative abundance of the various species may change substantially because of their different tolerances. Some collembolan species are independent of soil metal concentrations, and others are resistant or susceptible. The various mechanisms suggested to account for this (Tranvik and Eijsackers 1989; Tranvik *et al*. 1993) include reduced water-holding capacity of the soil, elimination of metal-sensitive fungi used for food, avoidance of metal-contaminated sites, and selection for metal resistance. In polluted soils near a brass mill in Sweden, one such springtail (*Folsomia fimetarioides*) thrived because it preferred metal-tolerant fungi and avoided those with enhanced metal concentrations.

Earthworms

Earthworms, also, can accumulate heavy metals as well as some pesticides and other organic chemicals which might affect their growth, mortality or reproductive capacity. However, and more generally, the impact of pollution in ecological terms is at the population (rather than at the individual) level (Moriarty 1983), with change in numbers being the most commonly assessed parameter of impact. Earthworms can be valuable as an assessment tool for toxic or hazardous wastes in soil. As an essentially cosmopolitan animal group, they are associated with virtually all applications of soil chemicals and vegetation changes that accompany agricultural establishment. They are regarded as among the most suitable indicator animals in testing for pollution by soil chemicals (Callaghan 1988), and have thus been selected as a 'key indicator organism' for ecotoxicological testing of industrial chemicals by the European Economic Community, the Organization for Economic Cooperation and Development, the Food and Agriculture Organization of the United Nations, and a number of national bodies of various countries engaged in pesticide registration and pollution monitoring. Protocols for assay, discussed by Goats and Edwards (1988), include examining responses to toxicants in natural soil, in silica paste on glass balls, by immersion in solution at various concentrations, by forced feeding tests, and by contact on filter paper. Short-term laboratory trials and longer-duration field testing are both feasible. Thus, in Britain, the pesticide chlordane (at 10 kg/ha–1) was toxic to two common earthworm species, *Lumbricus terrestris* and *Allolobophora longa*, for as long as 6 months on soil of low organic content, but its effect lasted for only about a month on soils of high organic content. Laboratory tests are needed for widespread use in such contexts, notwithstanding the substantial biological insight that can come only from field trials. Generally, though, laboratory tests on pesticide effects are of relatively short duration. Only 5 per cent ($n = 60$) of laboratory experiments using earthworms lasted for more than 6 months, compared with 68 per cent ($n = 63$) of field experiments (Lofs-Holmin and Bostrom 1988).

The ramifications of many such tests on pollution susceptibility are important in environmental management. Rhett *et al.* (1988) discussed several cases of using the worm *Eisenia foetida* as a monitor of contaminants in material dredged from water bodies, for which evaluation is needed (under the US Clean Water Act) before it can be discharged, together with assessment of the effects of dispersal on later contaminant concentration. *Eisenia* was considered to be a valuable monitoring agent, as it showed great facility to accumulate various heavy metals, polychlorinated biphenyls (PCBs), and polynuclear aromatic hydrocarbons (PAHs), and deposition of such chemicals in dredged soils used later in landscaping (or other) will inevitably be passed into other soil animals colonizing it. This examination thereby provides a 'first-step' evaluation of the mobility of particular contaminants which might have the potential to penetrate further into biological communities at a later stage.

Freshwater invertebrates

The responses of some invertebrate groups to pollution are difficult to assess simply because their 'normal' ecology and responses to normally fluctuating environments are not known in sufficient detail. This point was emphasized for freshwater sponges by Harris (1974). Mud (siltation) has long been regarded as the greatest threat to many freshwater sponges, and most taxa are indeed affected adversely by it. However, a number of sponges, including species of *Ephydatia*, are siltation-resistant. In

part this is because they grow in areas of low light intensity and/or on the undersides of their substrates. In Australian habitats subjected to severe annual flooding (and associated high silt loads), most resident spongillids are adapted well to these conditions, and suffer no perceptible harm (Racek 1969). Likewise, despite the general assumption of elimination of sponges from organically polluted waters, a number grow well in highly polluted regimes and sometimes even 'prefer' these. The pollution may well lead to a decreased number of sponge taxa, leaving only a few 'generalists', the common scenario for many invertebrates. Harrison (1974) noted the possibility of malformation of sponges caused by toxic pollution, and a number of early accounts were ambiguous over whether this occurred. For example, in early studies he cited, deformity of the axial canals of *Trochospongilla leidyi* growing in iron water-pipes, occurred. A similar malformation occurred in *Ephydatia fluvialis* in industry-polluted waters—but many other individuals of the latter were normal and the supposed deformities may have been a normal response to non-polluting environmental variables. This, again, indicates the difficulty of clear interpretation from incomplete knowledge of invertebrate biology. Some sponges may also survive short exposure to oils.

Total elimination of invertebrate groups from polluted waters has been reported—as for planarians from stretches of the North Platte River (Wyoming) polluted by the effluents of oil refineries (Neel 1953); planarians remained common above the pollution source and after 110 miles (*c.* 175 km) below it. Turbellaria were absent from waters polluted by lead mine effluents in part of Wales (Carpenter 1924) and by copper mines in Japan (Kawakatsu and Itô 1963). Kenk (1974) cited other planarian examples as well, but emphasized that, although such instances as these are clear, it is more difficult to determine the influences of pollution on planarian species replacement. Natural succession of species occurs in many water bodies, and this might be confused with pollution effects in leading to change in community composition. Planarians also exemplify the wide interspecific variability in tolerance to various pollutants that characterizes any large aquatic invertebrate group. Some (such as the European *Crenobia alpina* and *Dugesia gonocephala*) can withstand only mild pollution, and pollution tolerances are often correlated with temperate tolerances, so that eurythermic species are less sensitive to pollution than stenothermic ones. The kind of pollution is also often critical: degradable synthetic detergents in municipal sewage may have little effect on planarians (England: Hynes and Roberts 1962), as did dichloro diphenyl trichloro ethane (DDT) (at 1 p.p.m.) in a stream in Uganda (Hynes and Williams 1962), although many metallic salts and other pesticides are highly toxic. Some planarians may be more resistant than some other invertebrates: Hynes (1961) found that *Crenobia alpina* and *Polycelis felina* were not affected by overflow from a sheep-dip containing benzene hexachloride (BHC) residues, although amphipods and some insect larvae were almost completely eliminated.

Likewise, freshwater oligochaetes tend to be more sensitive to pesticide pollution than are many arthropods, but less tolerant of heavy metals (Brinkhurst and Cook 1974). Their variability in pollution tolerance was demonstrated by a survey of Tubificidae in the River Derwent, England, by Brinkhurst and Jamieson (1971). In 1958, when the river contained sewage effluents exceeding safe limits, only two species of oligochaetes (*T. tubifex, Limnodrilus hoffmeisteri*) were present in a series of samples. After a sewage disposal works had been installed, samples from the same stations in 1959–62 yielded a total of 12 species (6–9 at each of the four stations). 'Clean' water samples there, and in the Great Lakes of North America (study summarized by Brinkhurst and Cook 1974), still contained the two pollution-tolerant taxa noted above—so they are not truly 'indicator species', as has sometimes been suggested, but merely generalists in their ability to withstand pollution. This context is of much wider relevance.

Several species of leeches may have a similar wide tolerance (Sawyer 1974). These may feed on other aquatic oligochaetes and may be responding to the enriched food supply of superabundant tubificids (including *T. tubifex* and *L. hoffmeisteri*, as above) and other tolerant invertebrate prey. In common with many other oligochaetes, most leeches are absent from metal-polluted waters, but have an unusually high tolerance to DDT, with an LC_{50} greater than 100 p.p.m. Resistance of some species is based on their ability to absorb DDT and, sometimes, to dehydrochlorinate it to 2,2-bis-(parachlorophenyl)-1-1-dichloroethylene (DDE) (*Hirudo nipponica*: Kimura *et al*. 1967). The amphibious *Haemopsis sanguisuba* will climb out of contaminated water (Jones 1938).

In contrast to some of these groups, it appears that no species of freshwater ectoprocts are indicators of pollution, except by being absent (Bushnell 1974): although some species will tolerate moderate levels of pollution, they are generally rather sensitive to disturbance of any kind.

Limited eutrophication may help some molluscs (mussels, in particular) by increasing the available food supply, although harmful effects occur with increased levels of organic contamination. Mine drainage, predominantly of acid waters, can have very drastic effects on mussels (Fuller 1974) and gastropod snails (Harman 1974), but many molluscs can accumulate pesticides, as noted earlier for oligochaetes. Such materials can be concentrated in flesh but also shells, so that bottom sediments or 'subfossil' material can sometimes yield data on historical pollution levels.

Problems of acidification of water bodies, resulting from acid rains and runoff, have received considerable attention. Plastic Lake, Ontario, has been subjected to such acidification, especially from melting snow and other atmospheric input, so that its pH decreased from 5.8 to 5.6 over 6 years. The resident population of the crayfish *Orconectes virilis* became extinct during that period. The direct effects of acidification may have been augmented by increased predation by fish, resulting from the soft exoskeletons of the crayfish (France and Collins 1993).

Several studies have attempted to simulate the effects of acid precipitation by adding acid to streams or ponds and examining the effects on invertebrates. Hall *et al*. (1980) maintained a stream at Hubbard Brook (New Hampshire), normally of ambient pH greater than 5.4, at pH 4 from April to September 1977. The acidification led to decreased invertebrate species diversity, increased representation of community dominants (some caddis flies and stoneflies), and decreased complexity of food webs. Such 'simplification' by removal of many ecologically sensitive species is probably a very common effect of chemical pollutants.

Many pollution events in fresh water, especially in running waters, fall into the 'pulse' category of perturbation. It is usually by no means clear how long their effects may last—in extreme cases, far longer than the initial disturbance itself. Injection of the pesticide permethrin into an Ontario stream led to a dramatic but transient increase in macroinvertebrate drift (Kreutzweiser and Sibley 1991). The drift response was very rapid, with a wide range of taxa involved. The increase of Trichoptera was small. Overall drift densities in this instantly high-pollution regime rose to up to 4000–5600 times pretreatment levels, with a significant short-term reduction in benthos. Peak drifts subsided rapidly, and by 36 hours after treatment the densities at all sites measured had returned to control/pretreatment levels.

Multispecies samples in drift provide the opportunity, under such controlled application conditions, to examine the relative susceptibility of the taxa present to the particular pesticide used. The taxa showing the greatest sudden increase in drift are those with highest sensitivity, and ranking in order of increased abundance is broadly equivalent to ranking sensitivity or susceptibility.

It is widely assumed that recovery from pulse disturbances is rapid for lotic invertebrates (Palmer et al. 1992), though most cases documented are of scouring or flooding rather than of chemical pollution (Resh et al. 1988; Reice et al. 1990). Many attempts have been made to identify the reasons for this rapid return to equilibrium, commonly invoking consideration of the substrate as a refuge in which the organisms can shelter during a physical disturbance ('hyporheic refuge hypothesis') and/or drift and migration of the fauna to recolonize the disrupted region. This may assume that the animals 'dislodged' and displaced during the disturbance are not killed but move back later to the original area. The phenomenon of drift is also anomalous, because there is then a clear need for upstream migration

to avoid depletion of the populations there. A combination of factors is likely to be involved in most instances.

In contrast to pulse disturbances, 'press' disturbances reflect long-term pressures, such as occur frequently from industry, where effluents, such as heated water from power stations, may be released more or less continually. Effluents from abandoned copper mines in Japan (of which there are about 6000) led to moderate pollution with copper and zinc in several rivers studied by Hatekeyama et al. (1991), but negligible organic pollution because of low human settlement in the area. The insect community differed clearly between 'clean', moderately polluted, and non-polluted sites, with decreased numbers of species and lower diversity indices at sites where copper concentration was higher than 15µg/l. Metal-tolerant species included the mayfly, Baetis thermicus, which can concentrate high levels of heavy metals (Hatekeyama et al. 1991).

A second example involves 'thermal pollution' resulting from power stations using surface water for cooling. Wulfhorst (1991) commented that 'Every river ecosystem seems to react differently to thermal pollution', and generalizations do indeed seem difficult to find, probably because of the different (and, commonly, unknown) pollution levels compounding thermal effects and the different communities inevitably present in each river system. In Wulfhorst's (1991) study on the River Schwalm, Germany, the number of invertebrate taxa was only slightly less at thermally affected stations than the maximum number recorded above the power plant. Diversity and evenness were higher above the plant. Whereas three species of Gammarus all decreased downstream, the abundance of another amphipod, Asellus aquaticus, increased, as it was apparently 'stimulated' by thermal pollution. Similarly, particular species of Trichoptera either increased or decreased, but the relative abundance of some other taxa could be explained more adequately by changes in velocity or substrate type than by direct attribution to water temperature or quality.

Permanent changes to stream or river fauna can accrue through processes such as urban development, which have the capacity to alter dramatically watercourses that drain from such altered catchments. Local hydrology may be changed substantially, in addition to massive changes in water quality: for example, by deposition of inadequately treated sewage waters. A series of North American cases discussed by Jones and Clark (1987) showed:

(1) declines of pollution-sensitive invertebrate groups and increases in some others (such as particular oligochaetes);

(2) declines in the overall species richness from above to below urbanized areas; and

(3) changes in community balance to domination by few taxa.

Even in the absence of point-source discharges, watershed urbanization has major pollution impacts on benthic communities.

The upper Tennessee River (USA) has been subject to substantial change from industrial demands, so that permanent alteration in water quality and siltation has led to the endangerment of several freshwater mussel species. Compounding these effects, the exotic Asiatic clam (*Corbicula fluminea*) has become widespread since the mid-1970s, and is a potential competitor with native species. However, it is the major prey item of muskrats in parts of Virginia (Neves and Odom 1989). Muskrat predation has had little effect on mussel populations in the past, but now seems to be having much more detrimental influence on small and isolated populations of endangered mussels, such as shiny pigtoes (*Fusconaia cor*)—at the least, muskrat predation is retarding species recovery, and may be contributing to further declines. A necessary facet of management for mussel recovery may be removal of muskrats, particularly at sites regarded as refugia for the several endangered mussel species.

Pesticides as pollutants

The topic of 'pesticides as pollutants' merits further discussion here. Although they are classical pollutants as alluded to above, they differ from most other environmental chemicals in intent of use because they are designed to kill invertebrates and other biota, rather than being by-products of other purposes, and their side-effects are a cause of factual and emotional debate (Moore 1969, 1987). World-wide pesticide production increased massively in the decades following the end of the Second World War as part of a new phase of industrial and agricultural expansion, and the early high level of reliance on uncontrolled use of many potent chemicals with acute or chronic effects has now given way in the developed world to their regulated use in broader integrated pest management. However, elsewhere, their use still gives major causes for concern. The examples noted here, all from countries where pesticide use is strictly controlled, indicate a range of concerns.

'Codes' for careful application of pesticides to minimize side-effects (these being, mainly, death of non-target organisms or spread into natural communities, sometimes by concentration in food chains) are widespread. There is general opinion (Newsom 1967; Pyle *et al*. 1981) that most normally applied insecticides may cause only temporary disruption to other insect populations. The potential clearly exists for more damaging scenarios, especially for rare or sensitive taxa living close to agricultural or forestry areas subjected to intensive pesticide usage, or in areas where little or no regulation of their use can be enforced, or there are no alternative strategies for control of important pests.

In southern England, Rands and Sotherton (1986) compared the abundance of butterflies on an area of arable farmland subject to the usual array of cereal pesticides (including grass and broad-leaved weed herbicides, fungicides, and insecticides) with that on fields with unsprayed perimeters or headlands 6 m wide. Species richness over the 6 m strip seen in transects differed little (21 species on the unsprayed area, 17 on the sprayed area, total 22 species), but abundance differed significantly (868 individuals on unsprayed, 297 on sprayed, $P < 0.001$), leading to a suggestion that pesticide use may reduce butterfly numbers on arable farmland. Weed populations tend to be higher on unsprayed headlands, and many are larval foodplants or nectar sources for adult butterflies. More generally, such unsprayed headlands are associated with a greater cereal arthropod abundance (Rands 1985; Sotherton *et al*. 1989; Davis *et al*. 1991).

In Canada, non-persistent insecticides are used to control the major forestry pest caterpillar, spruce budworm (*Choristoneura fumiferana*) on conifers. Effects of aerial spraying on non-target organisms were estimated by comparing their 'activity' before and after application of aminocarb (4–dimethyl-amino-*m*-tolyl-methylcarbonate) in Quebec (Bracker and Bider 1982), using track marks on sand transects as an index. Inspection of one of the sites soon after spraying revealed large numbers of dead midges (Chironomidae), beetles, spiders, and harvestmen. Some differences in activity profile of invertebrates (caterpillars) in relation to control sites were indeed detected in this study, but (as a whole) arthropod activity seemed not to change significantly, and the long-term impact was considered likely to be small.

Likewise, contamination of water bodies by pesticides applied aerially over adjacent forests may only be transitory. For 1- (4-chlorophenyl)-3-(2, 6-difluorobenzoyl)urea (diflubenzuron) in Ontario, Sundaram *et al*. (1991) found significant reductions in 'waterfleas' (Cladocera) and some macroinvertebrates, but these resulted from direct overspraying in 2–3 months. Effects on copepods were less severe than on cladocerans.

However, small or vulnerable invertebrate populations could clearly be at risk from irresponsible insecticide application or from accident, and long-term or intensive application regimes might contribute to faunal change elsewhere, especially in localized or small habitats. The risk is likely to increase for species of conservation significance which are taxonomically related to the target pest (cf. biological control, p. 18). Thus, on Hawaii, the 'Tri-fly Project' was proposed to eradicate three species of exotic fruit flies from the archipelago by extensive aerial dropping of insecticide-treated baits to which the flies are attracted. Hawaii's vast array of endemic *Drosophila* flies are of massive evolutionary significance (see Carson and Kaneshiro 1976), and many species are

highly restricted in distribution. Most of the spectacular 'picture-winged' species each occur only on one island. The baiting proposal was regarded by many biologists as unacceptable because of the risk to these native species.

Wiest's sphinx moth (*Euproserpinus wiesti*) was considered to be one of the rarest North American hawkmoths (Sphingidae), and was known at one time only from a single locality in Colorado. As well as being sought ardently by collectors (p. 66), the moth had been threatened by insecticides applied aerially to control pest grasshoppers. In 1980, despite attempts to avoid the small (*c*. 600 m long) habitat, drift of malathion caused substantial mortality (Wells *et al*. 1983). Other grasshopper outbreaks could result in further inadvertent spraying of the area. Alternative grasshopper control methods (such as biological control by protozoans) may intervene to counter this threat. The moth is now known to occur elsewhere as well, but many invertebrates are indeed as vulnerable as the sphinx was, until recently, believed to be.

Poison baits applied to control pest vertebrates can harm invertebrates, but for some of these very little information is available and the effects are commonly disregarded. Compound 1080, sodium monofluoroacetate, is used widely for mammal control in New Zealand. Nine orders of invertebrates (Arthropoda, Mollusca) are known to be affected by 1080 (Notman 1989), either by eating bait, eating carcasses of poisoned animals, or by coming into contact with regurgitations or excretions of living poisoned vertebrates. In addition, sodium monofluoroacetate leached into soil may be taken up by plants, where it could have a systemic effect on herbivorous insect populations. As with many other pesticides, caution may be needed in applying such baits in areas known to support threatened invertebrates or protected species—in New Zealand, the three protected species of flax snails (*Placostylus*) may merit monitoring for any such effect.

EFFECTS OF EXOTIC SPECIES

The deliberate or inadvertent introduction of species from other parts of the world, or the natural invasion or colonization of a new area by such species, is commonplace, and constitutes a biological 'complication' to the receiving environment. These species are termed 'exotic', in the sense that they are alien to, and did not evolve in, the communities and areas that they come to occupy, and are often likened to a form of 'biological pollution'. Many exotic taxa are ecologically aggressive invaders whose effects in a 'new' area cannot be predicted reliably in advance, and which may adversely affect ecologically sensitive (often specialized and endemic) taxa when they arrive. Their collective threats to such invertebrates are viewed as secondary in importance only to direct habitat destruction.

The phenomenon of invasion by exotic species is not new: organisms have always been transported by people, and many historical introductions in the northern hemisphere (such as the honey-bee and rabbit in Britain, both introduced by the Romans) are accepted as typical members of the British fauna. In general, many such introductions by human agency are more recent in the southern hemisphere: in Australia, for example, over only around 200 years of European settlement. There, and on many islands, the ecological effects of exotic species are often more conspicuous because they can be seen against the background of relatively natural environments and biota. Introductions, purposeful or not, have increased markedly in number and extent with increased human traffic and commerce, and anthropogenic changes to newly-settled environments have facilitated the establishment of those many species that arrive naturally but which depend on other exotic species to sustain them. About 120 of Australia's 160 or so aphid species, for example, are exotic, and most of these could not invade before their foodplants (largely agricultural crops and exotic pasture grasses) were also introduced. It is rather unusual to have such a high proportion of exotic species within a given invertebrate group, but many formerly localized animal and plant species are now almost cosmopolitan, foreshadowing the disturbing trend foreseen by Elton (1958) that the world's distinctive regional biotas will become ever more similar and uniform as exotic species increase in diversity and effect, removing or causing extinction of many specialized biota. Few, if any, parts of the world remain unaffected and, whereas the effects of exotic plant weeds or large mammals may be dramatic and widespread, those of invasive invertebrates can also be pervasive and far-reaching. Such invaders are usually extraordinarily difficult to eradicate or contain once they have gained entry to a new region. They can spread rapidly and come to dominate their 'new' communities by superior competitive ability, with different taxa varying widely in this. In a comparison of three invasive sessile species in the Gulf of Maine (Berman *et al.* 1992), the tunicate *Styela clava* (probably originating from Japan) could be a competitive dominant in places, the tunicate *Botryloides diegensis* rapidly became dominant but this dominance was short lived, and the European ectoproct, *Membranipora membranacea*, overgrew native epiphytes to become a dominant species living on kelp. A recent marine arrival in Australia, the starfish *Asterias amurensis*, first detected in 1992, is already present in vast numbers in Tasmania, as a major alien predator, particularly of molluscs.

Aquatic invasions

The examples and contexts discussed here show some of the scenarios engendered by aquatic exotic species relevant to conservation of invertebrate species and communities. The list is by no means complete, and any of the substantial number of texts on biology of colonizing species (for recent examples, see P. A. Parsons 1983; Mooney and Drake 1986; Drake *et al.* 1989; Di Castri 1990) will reveal others.

One of the most thoroughly appraised marine biotic interchanges is that between the Red Sea and the Mediterranean since the opening of the Suez Canal in 1869. In common with other marine migrations, the general pattern is one of barrier removal facilitating movement from the richer to the poorer biota (Vermeij 1978). As well as about 30 fish species, some 20 decapod crustaceans and 40 molluscs (representing 26 families) have invaded the Mediterranean via the canal.

The Mediterranean immigrants include several active predatory portunid crabs, and a number of stress-tolerant molluscs, which are presumed to have tolerated the increased salinity of the Suez Canal on their way. Ecologically, many of the invaders are associated with boulders, live in seagrass beds, or have planktonic larvae.

More than a third of the 136 alien biota of the Great Lakes have arrived during the past 30 years or so, since the opening of the St. Lawrence Seaway allowed direct access by ships. The modes of arrival include:

(1) deliberate introduction as food species, often with 'contaminants', unless the stock is screened carefully for stowaway organisms;

(2) as hull-fouling taxa on ships;

(3) as shell-foulers on oysters and the like; or

(4) in ballast water, as in the above cases.

It is often extremely difficult to determine which mode has been the avenue for any given taxon. Further, Carlton (1985, 1989) emphasized the difficulty of actually determining whether an introduction had really occurred, because of poor knowledge of the natural distribution and taxonomy of some of the invertebrate groups involved, and because most biological surveys commenced well after many taxa had been transported globally. Several marine crustaceans and molluscs introduced to Australian waters are probable ballast-water species (Hutchings *et al.* 1987), and these authors enlarged on the problems of interpreting modes of arrival. They noted the following as possible ways of dispersal: drilling platforms, packing crates, discarded bait wrappings, wet fishing nets, bait wells, and seawater-intake pipes. The range of organisms involved is exemplified in Table 3.4, which also emphasizes the widespread uncertainties over origins and arrival dates for marine introductions, to many parts of the world.

1. The zebra mussel, *Dreissenia polymorpha*, is one of many marine species which have been transported in ballast water in ships, whence it was introduced from Europe (more specifically, from the Black Sea area) to near Detroit (Lake St. Clair) in 1985 or early 1986. After only 5 years, Bederman (1991) could claim that this accidental introduction to the North American Great Lakes had 'resulted in staggering monetary costs and incalculable environmental impact on that region', although the effects are not yet defined fully. The mussel is a notorious bio-fouler and has no natural predators in the Great Lakes. Cost of unfouling intake and outfalls of water treatment plants and power stations are anticipated to exceed US$5 billion during the next decade (Bederman 1991); interruptions to power and water supplies has occurred already; the smell of mussels may deter tourism, and piles of shells on beaches affect their amenity use. Zebra mussels have overrun the spawning reefs of sporting fish, and the contiguity of the Great Lakes to major American river systems suggests that it will spread rapidly to occupy large areas of the continent.

Effects on native invertebrates have not been documented fully, but are likely to be substantial. Native clams and crayfish can be killed by suffocation.

A second, related, species of mussel (termed 'quagga mussel') was detected in the Great Lakes in 1990. This is also from the Black Sea region.

2. In a related case, the interactions of native and introduced species are clearer. A European waterflea (Cladocera: *Bythotrephes cederstroemi*) is believed to have been brought to Lake Huron in the ballast of a Soviet tanker in 1984. It has now spread to the other Great Lakes, and native *Daphnia* (important as phytoplankton feeders at the base of the lakes' foodwebs) have decreased due to predation by *Bythotrephes*.

One important consequence of transferring invertebrate species from one region to another is that they might carry disease. Crayfish plague (p. 134) is one important example, and two other examples, one of which is difficult to interpret fully, are noted here.

Table 3.4 Marine invertebrate taxa introduced to Australian waters to 1986 (from Hutchings *et al*. 1987)

Taxon	Possible date	Possible origin	Possible method
Bryozoa (5 spp.)	post-1889 (1sp.), 1940s (1), 1950 (3)	Atlantic (3 spp.), Mexico (1), Japan (1)	Ships' hulls
Coelenterata Hydrozoa (1sp.)	1918 or before	Northern hemisphere	Ship's hulls
Annelida Polychaeta (5 spp.)	1885 (1 or 2), 1930s (1), 1970s (2 or 3)	Europe (2), India (1), Japan or Eastern Pacific (2)	Ballast or hulls (4), uncertain (1)
Mollusca Gastropoda (8 spp.)	Unknown (1), 1920s (2 or 3), 1950s (2), 1970s (2)	Japan (2 or 3), New Zealand (2 or 3), California (1), South Africa (1 or 2), Europe (?1 or 2)	Ballast or hulls (1), ships' hulls (5), with oysters (1), unknown (1)
Bivalvia (7 spp.)	?1876 (1), 1880s (1), 1940s (1), 1950–60s (2), 1970s (1)	New Zealand (4), Japan (1), Pacific coast of Asia (2)	Ballast (2), with oysters (3), deliberate (2)
Polyplacophora (1 sp.)	?1910	New Zealand	With oysters
Brachiopoda (1 sp.)	?1901	New Zealand	With oysters
Arthropoda Cirripedia (4 spp.)	1940s (3), 1981 (1)	Japan (1), South Africa (1), Atlantic (1), unknown (1)	Ships' hulls
Mysidacea (1 sp.)	1977	Japan	Ballast
Isopoda (6 spp.)	1920s (1 or 2), 1970s–80s (4)	USA (?3), Indian Ocean (1), New Zealand or Chile (1), unknown (1)	Ships' hulls (6)
Decapoda (4 spp.)	1880–1930 (1), 1900–1980s (1), 1920s (1), 1970s (1)	New Zealand (2), Europe and unknown (1), USA or Asia (1)	With oysters (1–2), ships' hulls (1–2), unknown (1–2)
Echinodermata (1 sp.)	1930s–50s	New Zealand	With oysters
Ascidiacea (3 spp.)	?1878 (1), 1960–70s (2)	North Atlantic (1), unknown (1), ?Europe or North Pacific (1)	Ships' hulls (3)

1. Oysters have long been transported between various parts of the world to found aquaculture industries. During the period 1966–77, large numbers of Pacific oysters (*Crassostrea gigas*) were introduced, as 'seed' and adults, to France from Japan and British Columbia, to replenish oyster stocks. This was necessary because of the disease-induced decline of *C. angulata* (Sinderman 1986). The oysters established so successfully that production along the French coast was more than 100 000 tonnes by the late 1970s. Among other organisms introduced during the oyster importation were a parasitic copepod, *Mytilicola orientalis*, and several Pacific seaweeds.

Crassostrea gigas introductions coincided with epizootic outbreaks in native oysters, not only *C. angulata* (which had been largely replaced by *C. gigas* in many areas by 1973: Comps *et al.* 1976), but also *Ostraea edulis* (the European flat oyster). Two protozoan parasites successively reduced these oyster populations dramatically. Sinderman (1986) commented that the succession of three major epizootics within a decade was unique in the history of oyster culture but, despite suggestions that agents imported in mass introductions of an exotic oyster might have participated in disease outbreaks in the native species, this relationship has not been proved.

2. In some other instances, more direct evidence of disease spread via host release is available (see also p. 124). The virus known as IHHNV (infectious hypodermal and haematopoietic necrosis virus) is a serious threat to the aquaculture of penaeid shrimps. It is presumed to be endemic to the west coast of Central and South America (Fig. 3.2) but has been introduced into aquaculture facilities in many parts of the world with infected shrimps. The possibility of infection of native shrimp populations by the introduced pathogen is of considerable concern, despite lack of evidence that this has happened yet, and strict quarantine (p. 123) of imports and selection of virus-free breeding stock is vital (Sinderman 1986).

Similar caveats apply, of course, to other ecosystems and to all kinds of introductions, including vertebrates. The latter may assume considerable significance as threats to invertebrates, particularly when higher-level consumers are involved. In freshwater ecosystems, fish have

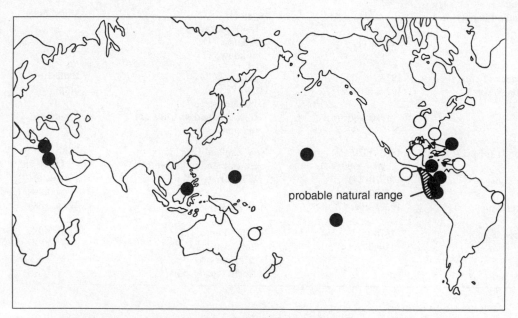

Fig. 3.2 Natural range and global distribution of IHHNV of penaeid shrimps as a result of transfers and introductions for aquaculture: closed circles, confirmed; open circles, suspected. (After Sinderman 1986.)

been distributed widely: as predators (biological control agents) of mosquitoes (*Gambusia affinis* is known as the 'mosquito fish'), recreation (trout, *Salmo* spp.) or food (*Tilapia*, carp) for example, and their ecological effects have rarely (if ever) been anticipated clearly.

Concerns over effects of exotic fish on Australian freshwater invertebrates exemplify a much broader invasion potential. Concern has arisen from fish in the first two of the above categories (Fletcher 1986), though effects of various other fish, such as aquarium discards, have not been appraised as fully. Many liberations of the polyphagous *Gambusia* were made in Australia from about 1925 onward (Wilson 1960) and its spread was fostered actively over several states by army malarial control units from around 1940 on. It has been implicated in the decline of aquatic insects other than mosquitos, although most such effects have been inferred merely from the vast numbers of mosquito fish present in small water bodies and knowledge that they are indeed generalist predators. Brown trout (*Salmo trutta*) in Australia have been associated more particularly with the decline of Tasmanian endemic syncarid crustaceans, *Anaspides*, following introduction to many lakes in Tasmania in the nineteenth century. *Anaspides tasmaniae* may now survive only where trout are absent or in low numbers (Williams 1981).

Trout have been implicated also in the decline of a local planorbiid snail (*Ancylastrum cumianganus*: Smith and Kershaw 1981) and in local elimination of Plecoptera and Trichoptera in Australia (Fletcher 1979). Trout feed mainly on benthic invertebrates and the availability of suitable prey was probably instrumental in assuring the success of early introductions to Australia (Tilzey 1977). Similar scenarios can be projected for New Zealand (Collier 1993), and elsewhere. In the isolated Hawaiian islands, apparent extinction of the endemic damselfly, *Megalagrion pacificum* on Oahu was probably due to predation by introduced fish (Zimmerman 1948), predominantly *Gambusia* and others imported to control mosquitoes.

Brine shrimps, *Artemia*, have been introduced into saline lakes of many parts of the world, with little historical regard for specific or intraspecific variations. Concern has been expressed that such introductions may foster competition with local *Artemia* and lead to loss of genetic diversity. Under experimental conditions, it has been shown that one strain may become extinct, and sexual strains may be able to outcompete parthenogenetic strains (Browne 1980). *Artemia* is exotic in Australia, where Geddes and Williams (1987) have expressed grave concern over its future effects on the diverse endemic salt lake brine shrimp genus, *Parartemia*. Generally, *Artemia* in Australia is restricted to coastal saltworks, but there are recent records from isolated inland sites.

Horwitz (1990a,b) emphasized the importance of another 'level' of 'exotic' introductions for certain freshwater crustaceans: that of more local translocations within a country of groups of native taxa whose systematics is poorly understood. He stated that 'translocation of Australian native species of freshwater crayfish has been effectively unrestrained within states, between states . . . for substantial periods in the past'. The problems Horwitz noted which could potentially arise from this practice are:

(1) introduction of diseases such as crayfish plague and porcelain disease;

(2) competitive interactions between introduced and native species;

(3) loss of unique combinations of characteristics through hybridization; and

(4) loss of unique combinations of epibiotic species.

Horwitz (1990a,b) urged a review of translocations of freshwater decapods in Australia, where freshwater crayfish are indeed more diverse than had long been suspected. *Cherax tenuimanus* ('marron'), for example, is a large commercially desirable species native to Western Australia but which has been translocated to many parts of Australia. Two possible subspecies are recognized, however, and Horwitz noted the possibility of these hybridizing intermittently to weaken the biological integrity of each. In general, genetic information on such cases is non-existent.

Regulation (p. 124) may be needed at local, as well as international, levels to help control such introductions.

Terrestrial invasions

Many adverse effects of exotic species in terrestrial communities have been documented extensively, and they tend to be more conspicuous than in aquatic systems. Again, deliberately introduced species and 'accidental' arrivals can be involved, as the following examples show.

Biological control

Classical biological control (CBC) of pest arthropods and plant weeds has long been an important component of pest management strategies, valuable in leading to reduced pesticide use on many crops and weeds. Criteria that allow a given biological control agent to be imported or introduced have gradually been refined from the initial historical emphasis on simply whether it would eat or attack the target organism. With that emphasis, a number of relatively polyphagous predators and parasitoids were indeed liberated without full appreciation of the possible adverse side-effects of any feeding on non-target species in their new environment. Thus, although CBC was heralded widely as 'environmentally safe', mainly in contrast to the 'unsafe' continued use of toxic pesticides, certain aspects of its use may constitute an environmental threat.

Howarth (1983, 1986, 1991) discussed, in particular, the effects of a multitude of biological control agents on the isolated and highly endemic Hawaiian biota—which may be far-reaching. Zimmerman (1948) believed that polyphagous parasitoids introduced to control pest moths had 'resulted in the wholesale slaughter and near or complete extermination of countless species' of native Lepidoptera. Direct evidence is sparse, because massive habitat destruction occurred over the same period, but Gagné and Howarth (1984) also implicated exotic ants and biological control agents in the extinction of 16 moth species. Reduction of native caterpillars may have led directly to rarity or extinction of native predators (such as *Odynerus* wasps: Zimmerman 1948), and also contributed to the

decline of insectivorous birds through reduction of food supply (cf. p. 63).

Samways (1988) emphasized that, because the target for CBC is usually itself exotic, it is not itself likely to be a conservation target. Quite the reverse, in fact! Any switch by an introduced natural enemy to another host or prey species has most commonly been to a taxon related to the target species. Referring to arthropod agents, Samways (1988) noted that instances where categorical harm to native fauna had occurred were based on very limited evidence, despite several allegations of the kind noted above. Indeed, CBC might *help* conservation programmes in some instances, by reducing infestations of exotic pests in protected areas.

In general, biological control agents constitute a very small proportion of the exotic insects of any local biota, but the responsibilities of assuring the 'safety' of such deliberate introductions are well recognized, and protocols for assessing safety are being refined progressively. These have been developed rather more rigorously in the past for herbivores introduced to control weeds (Harris 1973; Wapshere 1974) than for higher-level consumers to control arthropods and other invertebrate pests.

Generalist predators, in particular, are undesirable introductions (cf. p. 63) and, ideally, biological control agents should be entirely specific to the target, a situation which is both rare and difficult to establish beforehand.

One undoubted case of a predator introduced deliberately to control an exotic invading species, which has caused massive conservation concern because of its unanticipated side-effects, is the snail *Euglandina rosea* on various Pacific islands. It was introduced to control the giant African snail, *Achatina fulica*. *Euglandina* has had devastating effects on endemic terrestrial snail faunas and, as it has been shown recently to have the ability to capture and eat aquatic molluscs as well (Kinzie 1992), its effects are likely to be even more drastic than documented up to now. *Achatina fulica* has been dispersed widely by people (Fig. 3.3) as a source of food, but also became a serious agricultural pest (Mead 1961, 1979) because it fed on crops and increased

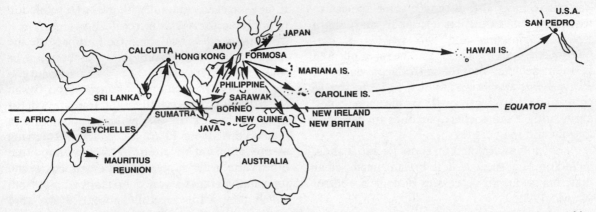

Fig. 3.3 The giant African snail, *Achatina*, dispersal by people. Arrows show the broad dispersal patterns from Africa documented by 1950. (Anon 1950.)

rapidly to reach massive numbers. The impact of *Euglandina* on *Achatina* has been doubtful in many places, and its effects on other molluscs (exemplified by the two cases below) have led to recommendations that its introduction elsewhere should be prohibited entirely.

Partula on Moorea

Moorea, near Tahiti (Society Islands), is a volcanic island about 12 km in diameter, and supported a group of endemic species of *Partula* land snails of unusually high interest in studies of variation and speciation. As many as four of the seven species could coexist without interbreeding (Clarke and Murray 1969; Murray and Clarke 1980). *Euglandina* was released on Moorea, with official permission, in 1977. By 1980 it had spread over nearly a third of the island (Fig. 3.4). Field surveys in 1982 (Clarke *et al.* 1984) showed several formerly abundant species of *Partula* to be absent, apparently having been eliminated entirely from areas where they were common in 1967 and 1980. *Partula aurantia*, formerly restricted to the north-east of Moorea (Fig. 3.4) was by then considered to be extinct in the wild, and Clarke and his colleagues predicted the times of extinction of all seven species on Moorea as, variously, 1983–87.

In surveys during 1987, no *Partula* could be found on Moorea (Murray *et al.* 1988). The genus seemed to be extinct there in the wild, although several species were then presumed

to be secure because of captive breeding stocks held in other parts of the world (p. 100).

Partula has also declined substantially and rapidly on Tahiti, where the decline was also associated with the spread of *Euglandina*, and on Huahine (where *Euglandina* had not been introduced) populations of *P. varia* and *P. rosea* continued to flourish despite being sought locally for trade in shell jewellery. On Guam, where

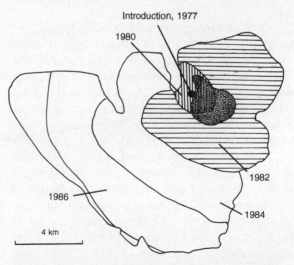

Fig. 3.4 *Partula* and *Euglandina* on Moorea. The spread of *Euglandina rosea*, from introduction to Moorea in 1977, by 1984 and beyond (predicted). The dense, shaded area was the range of one species of *Partula*, *P. aurantia*, which had become extinct in the wild by 1982. (After Clarke *et al.* 1984.)

Partulidae have also declined, some species to the point of extinction (Hopper and Smith 1992), *Euglandina* and two species of carnivorous *Gonaxis* snails have been implicated in their decline, but the introduced flatworm *Platydesmus manokwari* (which was effective in controlling the African snail) has also been noted abundantly in areas where partulids are declining rapidly and may eat them.

Achatina has declined on some Pacific islands, including Hawaii, but it is by no means clear that this is due to successful biological control (Cowie 1992).

Achatinellidae on Hawaii
The Achatinellidae are one of several diverse groups of land snails in Hawai'i (Solem 1990) which have declined recently. Solem's sobering estimate was that only about 25–35 per cent of the 1461 species of endemic land snails are still extant, with most of these likely to disappear within a few years. Achatinellidae are one of three families which have undergone 'catastrophic extinctions' there, and many species of *Achatinella* are included (Hadfield 1986). Early detection of this decline led to all endemic species of *Achatinella* being designated as 'endangered' on the US Federal Register (1981) (p. 115).

Many factors contributed to the decline, including habitat change, shell-collecting (p. 66), and exotic species such as ants and, possibly, the flatworm *Geoplana septemlineata*. Disappearance of a population of *A. mustelina* on Oahu coincided largely with invasion of the area (in the Waianae Range) by *Euglandina* (which was introduced to Hawaii in 1955–56). No *A. mustelina* could be found in 1979 (Hadfield and Mountain 1981), after *Euglandina* had been found there in 1977: Hadfield and Mountain concluded that *Euglandina* was the direct cause of extinction of that population. Although Hadfield (1986) regarded *Euglandina* as 'probably, the most serious modern predator of native Hawaiian snails', other mortality factors operate in areas where *Euglandina* is not present, and diseases of various sorts may also be involved.

In developing CBC methods and protocols, it is now agreed generally that vertebrates are undesirable for use as introduced agents. Much of the awareness has come from effects of vertebrates on other vertebrates (the brown tree snake, *Boiga irregularis*, has been responsible for a number of vertebrate extinctions on Guam (Nafus 1993); foxes and other predators have been implicated in loss of small mammals in Australia: (Myers 1986), for example), as well as predation on invertebrates. However, some other vertebrates have been dispersed accidently; for example, rats occur now on many islands from which they were naturally absent. Rats were noted by Hadfield (1986) and earlier workers as a factor in the decline of the Hawaiian snails discussed earlier, and concern over rats on islands is widespread in relation to the well-being of large (especially, ground-dwelling) invertebrates. They have been implicated in the disappearance of large orthopteroid insects, such as the Lord Howe Island stick insect (*Dryococelus montrouzieri*) and various 'wetas' (sometimes termed 'insect mice') (Howarth and Ramsay 1991) on islands off New Zealand (p. 140), for example. The cane toad, *Bufo marinus*, is a well-known example of a species dispersed in part deliberately for biological control, and has been the subject of considerable accusation over harming invertebrates in places where it has been liberated (discussed by Niven 1988).

Social Hymenoptera
Ants, wasps, and (more rarely) bees have been regarded commonly as among the most invasive exotic arthropods in terrestrial communities, and cases for adverse effects on native invertebrates are commonplace. These insects tend to attract attention because of human fear, so that concern over their invasions usually has a 'public health' impetus as well as ecological considerations. The two examples given indicate the kinds of problems that can ensue.

Imported fire ants, *Solenopsis* spp., in the United States
Two species of *Solenopsis* (the 'red imported fire ant', *S. invicta*, and the 'black imported fire ant, *S. richteri*), were introduced into the USA early this century and have spread widely across the southern States (Lofgren 1986; Vinson and

Greenberg 1986). Worker ants are particularly aggressive. They attack and sting any animal that disturbs their nest, and foragers sting people readily, frequently leading to allergic reactions. The ants are polyphagous and sometimes cause extensive damage to crops. Their generalist predatory habits have led to many reports of harmful effects through feeding on beneficial insects, such as carabid and staphylinid beetles, as well as beneficial effects by feeding on pest caterpillars and Homoptera. Likely effects on invertebrates have been overshadowed by reports of fire ant predation on vertebrates (such as quail chicks, ducklings, hatchling turtles, lizards, and snakes) with declines in some of these species attributed directly to ant predation. There is little doubt that effects on native invertebrates could be substantial, and the lack of detailed information on this reflects the very common scenario of ignoring invertebrates in charting the effects of invasive species.

The Argentine ant (*Iridomyrmex humilis*) and the big-headed ant (*Pheidole megacephala*) are among other invasive ants in various parts of the world, which are implicated strongly in the decline of native ant species through being aggressive competitors, and which probably also have adverse influences on other taxa.

Alien predatory ants, particularly the big-headed ant (*Pheidole megacephala*) and the long-legged ant (*Anoplolepis longipes*) are major threats to native fauna in Hawaii. They have been implicated strongly in exclusion of native tetragnathid spiders from native and disturbed forests, for example. In a series of laboratory trials in which six or seven species of *Tetragnatha* were caged with one or other ant species, the spiders were always killed (Gillespie and Reimer 1993). This was in contrast to a number of other species examined under similar conditions, and may be due to their larger size, harder exoskeleton, or (in *Oxyopes* sp.) ability to autotomize legs, allowing the spider to escape.

European wasps, *Vespula* spp., in New Zealand

Two species of European *Vespula* social wasps occur in New Zealand: *V. germanica* colonized in the 1940s and *V. vulgaris* in the late 1970s. The latter has replaced the earlier invader in some regions and is now the more abundant species in native *Nothofagus* forests (Thomas *et al*. 1990). *Vespula germanica* is still the more widespread species in New Zealand, and is the more abundant in many non-forested rural areas (Clapperton *et al*. 1989). Worker wasps forage actively for carbohydrate supplies and arthropod prey. Carbohydrate utilized in New Zealand *Nothofagus* forests is predominantly the honeydew of a scale insect, *Ultracoelostoma assimile*, and the wasps are the predominant consumers of this (Moller and Tilley 1989): they may compete advantageously for honeydew to the detriment of native insects, honey-bees and, even, endemic forest parrots (Beggs and Wilson 1991).

A study of the wasp's arthropod diet, undertaken by intercepting foragers returning to the nests with prey (Harris 1991) showed that both species capture many kinds of prey, including representatives of nine insect orders and some other arthropod groups. Extrapolation led to estimates (for *V. vulgaris*) that colonies in western South Island beech forests take an average of 1.4 kg of prey/ha, and in northern South Island this increased to around 8.1 kg/ha/year. Much of this is likely to be of native (mainly endemic) taxa, and raises concerns similar to those voiced for invasive Vespidae elsewhere—such as *Paravespula pensylvanica* on Maui, Hawaii (Gambino *et al*. 1990). Several species have been reported as pests by bee-keepers, indicating the likely high levels of impact in natural communities where such concentrated prey may not be available easily.

The above two species of *Vespula* have both been introduced to Australia, where *V. germanica* is the greater concern (Crosland 1991) and could potentially colonize much of eastern and southern Australia. Its environmental effects have not yet been assessed, but there seems no reason to doubt that they will be any less severe than in New Zealand.

One aspect of exotic species which is difficult to assess is the possible synergism between multiple species. A widespread aspect of CBC in the past was to introduce a spectrum of the natural enemies of a weed or pest insect in the hope that they would have an increased collective effect on the target species. With more than 2000 exotic arthropods, alone, in Hawaii (Howarth 1986), such interactions are indeed likely. Howarth cited the expansion in numbers of some exotic pest ants facilitated by the availability of numerous exotic honeydew-producing Homoptera and, more generally, demonstrated a number of ecological contexts where two or more 'aliens' may have their joint impact increased: alien prey species for polyphagous alien predators, pollination of alien plants (including weeds), dispersing alien plant propagules, and others that could increase their overall impact.

Many terrestrial exotic species are more or less synanthropic and remain closely associated with people, human habitation, or introduced resources and highly modified environments. The concern for conservation devolves primarily on those which invade natural ecosystems, as exemplified above, but it is very difficult (other than by the same species' record elsewhere) to predict whether this is likely to occur.

Invasion may well depend on local conditions at any time, including the human demand for domestic species to intrude into natural regions, or for those regions to be changed to accommodate such species—so that natural habitats are destroyed to facilitate agriculture, for example. More subtle consequences also occur. Clearing of *Eucalyptus* woodland for farming in southern Australia, for example, has reduced the nectar supplies available for exploitation by the substantial bee-keeping industry founded on the introduced honey-bee, *Apis mellifera*. There is now pressure for the industry to be allowed to exploit the nectar supplies available in protected areas (such as National Parks), with concerns that the bees could compete with the numerous native bee pollinators for a limited nectar resource. Current research programmes are helping to clarify the ecological roles of honey-bees in the Australian environment.

Extent of the threats

Eradication or containment of exotic species is advocated frequently, but is extraordinarily difficult to achieve; once an aggressive colonizer gains a foothold in a new environment, it may be virtually impossible to eliminate. Attempts to eradicate introduced invertebrates, even those which have no known side-effects, have rarely succeeded—especially when pesticides are used. The freshwater golden apple snail, *Pomacea canaliculata*, was introduced to Taiwan from Argentina as human food in 1979–80, thence to Japan in 1981 and later to several other countries in South-East Asia (Mochida 1991). It was cultured in ponds and sold fresh, bottled, or canned, but at present has little commercial value. However, it soon escaped from aquaculture and affected aquatic plants in open fields. The snail occupied more than 170 000 ha in Taiwan by 1986, and 16 000 ha in Japan and around 400 000 ha in the Philippines by 1989–by which time it had become an important pest of crops such as rice seedlings, taro, swamp cabbage, lotus, wild rice, water chestnuts, and others. Rice fields are the most serious concern.

In Taiwan, well over a million fingerling common carp, and more than half a million black carp were released in rice fields as potential snail predators in 1984 and 1986, and large quantities of molluscicides were used. The snail is proving difficult to control, but the above 'methods' are likely to affect non-target organisms as well. In the Philippines, *Pomacea* has been reported to replace a native relative, *Ampullaria luzonica*, and such coexisting species could clearly be affected adversely even more by non-specific control measures applied against *Pomacea*.

The effects of exotic species may sometimes be more extensive than we suspect, and in ways that are not obvious except with hindsight. Two examples of endemic mollusc radiations exemplify this further.

1. Small hygrobiid snails are one of the endemic radiations of Australian animals associated with permanent arid-zone springs, and their systematics and evolution are only now being appraised.

Ponder and Clark (1990) described 12 new species of *Jardinella* snails from Queensland springs, for example, and regarded 11 of these as endangered. The habitats of all of these 11 were on pastoral land, and they occurred in small numbers of springs, sometimes closely grouped, which can be degraded easily. Damage can occur from pastoral stock and from feral populations of (particularly) horses, camels, and donkeys. Direct damage from pastoralists providing for their stock can occur from damming, placing a well on or near the spring, or digging out. Indirect damage can occur by making artesian wells, so that natural artesian flow is reduced and eventually ceases. The latter has led to many wells in Queensland, and virtually all in New South Wales, already becoming extinct, and this trend is likely to continue unless removal of artesian water is regulated strictly.

2. The largest group of New Zealand land snails are the carnivorous species of *Powelliphanta* (Paryphantidae), with many local forms and controversial taxonomy: 10 species, 34 subspecies and four 'forms' were recognized by Meads *et al*. (1984), reflecting earlier classification. The snails may live for 40 years or more, and there may be 15 years between generations.

All species are of concern for conservation, and some are close to extinction. Most occur in native forests, a habitat type reduced to small remnants in some areas. Where forests are browsed by (exotic) ungulates, the litter can be disturbed and dried, and the ground cover can be removed, exposing the snails to predators. Meads *et al*. recorded some of the most severe predation levels in forests highly modified by a number of introduced mammals, including pigs, deer, goats, sheep, cattle and possums. Pigs probably have the greatest effect, because they root up the litter layer, also eat snails, and frequent scrub at the edges of forests, so that small remnant patches are particularly vulnerable because of their high edge : area ratio. Native predators (birds) also eat the snails, but exotic predators appear to be much more important, particularly when associated with habitat decline. Rat and pig control seems necessary for effective management for most *Powelliphanta* species.

OVEREXPLOITATION AND OVERCOLLECTING

'Overcollecting' is an emotional topic in conservation management, and has been implicated as a major threat to a number of invertebrate groups in two contexts, in particular:

(1) local overexploitation or excessive take of commercially desirable taxa not in themselves rare, such as food species and other commodity species;

(2) overcollecting of rare species, such as rare butterflies, molluscs, or corals, desired by collectors and which can fetch high prices.

Many of the latter category may occur in very restricted areas which can be systematically searched for every available specimen.

Possibilities for overcollection of invertebrates thus occur in many different contexts, with the above categories intergrading. Commonly cited contexts include: uncontrolled harvesting of seashore molluscs for food, or worms or tunicates for bait; of sponges for commerce; of many marine taxa for food; of reef taxa for the aquarium and tourist industries; of freshwater worms and insects for bait; of rare beetles, butterflies, and molluscs for sale to collectors. The scales of exploitation, reflecting many of the 'uses' of invertebrates discussed in Chapter 2 range from high-value food crops on which local industries are founded, to small-scale or low-return industry (bait fishing for sale to tourist fishermen), to individual collection of specimens for collectors and museums. Both local and international markets may be involved. Much collection for food or bait, for example, is typically local, but the former, particularly, can cater for export markets as well. Figure 3.5 shows the fate of a popular bait organism exploited by fisherman, hellgrammite larvae (Insecta: Megaloptera: *Corydalus cornutus*), taken from the New River area of West Virginia (Nielsen and Orth 1978), but estimates of the amount/numbers of invertebrates captured in most such enterprises are not available, and it is often very unclear just what effects 'normal' or 'usual' levels of collecting have on stocks of any reasonably common invertebrate taxon. Two species of intertidal polychaete worms in Maine comprise more than 90 per cent of the United States baitworm fisheries (Brown 1993). Take has been up to 400 000 lb/year of worms in the past, but this has declined substantially in recent years, probably in part due to increased scarcity resulting from overharvesting in the past. Similar concerns are expressed often and some form of 'overexploitation' is suggested commonly to be a reason for scarcity of a target species. These species are scarcely considered for practical conservation, but the lessons to be learned from practical management to sustain stocks are clearly relevant to rarer species—as are the causes of their declines.

For many groups, harvesting methods are now more efficient than ever before, so that even relatively recent historical underexploitation based on more traditional hunting methods may be countered. Demand for many commercial food species, in particular, is being satisfied increasingly by mariculture, aquaculture, and related 'controlled breeding' operations (p. 102), but this is not yet a viable option for others to guarantee their sustainability. However, it is often unclear whether overcollecting really does occur, despite evidence of persistent taking of specimens from some natural populations. Rarely, if ever, are

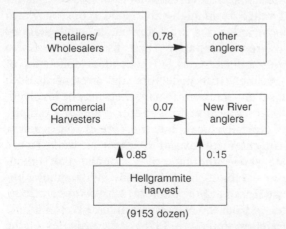

Fig. 3.5. *Corydalus* larvae and bait: exploitation of hellgrammites for fishing bait in the New River system, West Virginia. Percentage of trade flow shown. (After Nielsen and Orth 1988.)

sound population data available to confirm this as the primary cause of decline. Population studies on the red sea urchin, *Strongylocentrotus franciscanus*, showed that establishment of harvest refuges is needed to sustain catches for the long term by protecting adequate reproductive stocks (Quinn *et al.* 1993).

Hadfield (1986) and Solem (1990) regarded shell-collectors as among the most serious 'predators' of Hawaiian achatinelline snails, because they could directly seek out and exploit isolated populations. Kondo (1980) also attributed snail extinctions to overcollecting, and these authors documented instances of intensive collecting of many thousands of snails from single localities, often apparently removing all specimens encountered, and of collectors, accompanied by troops of native assistants, ransacking each valley.

Perhaps the most emotive arguments centre on such 'collectable' taxa rather than on commercial food species. Their appeal to collectors is often simply because of their rarity. In contrast to the snails mentioned above, among butterflies (despite a substantial trade) there are few unambiguous cases where collecting has clearly harmed the population or species concerned to the point of increasing vulnerability to extinction. Such considerations are of more than mere academic relevance, because the perceived dangers of overcollecting form the basis of much of the protective legislation (p. 114) which has been enacted over invertebrates. Taking of even a few individuals from small, vulnerable populations may indeed 'tip the balance' against survival, but it is more doubtful whether overcollecting is commonly the *main* cause of populations initially becoming vulnerable. But, in such circumstances, 'overcollecting' may occur with taking of only one or two individuals.

For the British large copper butterfly (*Lycaena dispar*, p. 100), a combination of increased collector interest, including that fostered by commercial dealers, *and* decreasing habitat assured its extinction, around the middle of the nineteenth century.

Three main categories of exploitation, reflecting different kinds of commercial demand, occur for butterflies (Collins and Morris 1985). These are important in illustrating the complexity involved in trying to consider general collecting effects for many other kinds of invertebrate. The categories are:

1. The 'high value/low volume' trade is of greatest concern for conservation. It involves the capture of very rare species for sale, individually and often illegally, to wealthy collectors. Unit price may be high: some birdwing butterflies have been advertised at up to US$7000/specimen, and the more elusive females tend to command higher prices than males. Such exploitation can indeed affect low-density and vulnerable populations in a very purposeful way; especially so as many such species occur in regions where collector-labour is cheap and abundant, and populations (even if the species involved is legally protected) are extraordinarily difficult to patrol effectively.

2. The 'low value/high volume' trade is that which supplies the vast range of 'souvenirs'—the various butterfly ornaments which abound in many hotels and airports in the tropics—and individually mounted or papered specimens of many of the more common species to collectors not seeking the greatest rarities. The value of this trade is vast and difficult to estimate, even within very broad limits: for Taiwan alone, estimates in the past have suggested a former annual value of US$2 million (Marshall 1982) to a staggering US$30 million (Owen 1971), with more than a thousand dealers involved. There is emphasis on larger 'showy' species, and in many instances only the wings are utilized in ornament-making, with the bodies being discarded or used for pig food.

It would seem inevitable that such massive and sustained exploitation would indeed harm natural populations, but the effects are by no means as clear, and some workers claim that *no* harm results. This, in part, is because much of the collecting is undertaken by use of attractant baits (such as carrion, dead fish, fermenting fruit), and only male butterflies are susceptible to these. Whereas many male butterflies are polygamous, many females mate only once and it is believed commonly that, because females are not captured by the above methods, they may indeed survive to breed without any diminution

of the effective populations in each generation. More data are needed on this controversial question.

3. The live trade to provide material for display in butterfly houses and similar exhibitions, which have increased in several parts of the world (Collins 1987a). Although increasingly specimens are being bred by the institutions involved, there is still substantial reliance on imported stock, mainly of large, showy, long-lived species, and with a high proportion supplied by dealers in South-East Asia. Much of this derives directly from wild-caught stock, exported mostly as pupae. Some local butterfly-farming or ranching operations (p. 102) alleviate pressure on wild populations somewhat.

The rational harvesting, or sustainable use of natural resources, is an important need of practical conservation, espoused by the World Conservation Strategy (IUCN, UNEP, WWF, 1980) and it is often difficult to determine when the transition from a reasonable level of exploitation to a threat takes place, because of natural variations in population numbers and in harvesting intensity. Drought and crop failure, for example, led to increased collecting intensity on food invertebrates from rocky shores in the Transkei (Hockey and Bosman 1986). Proximity to human settlement is also relevant, both for local food species (intertidal zones in Chile: Moreno *et al.* 1984) and for commercial operations such as offshore fisheries.

Some invertebrates have indeed declined by being collected for a particular purpose, such as food (some molluscs and crustaceans), medical use (the medicinal leech, *Hirudo medicinalis*, p. 133), or desirability as unusual pets (the red-kneed bird-eating spider, *Euathlus (Brachypelma) smithii*; p. 101). The latter, together with some hermit crabs and molluscs kept as pets, has some potential to become an invasive and undesirable exotic species if released (or if it escapes from captivity) outside its native Mexico. Other invertebrates are harvested for many reasons: horseshoe crabs (especially *Limulus polyphemus* from the Atlantic coast of North America) are exploited for biomedical research (relatively recently: Ruggieri 1976), for zoological teaching

Table 3.5 Reasons for overexploitation of invertebrates

Taxon	Main reasons[a]										
	1	2	3	4	5	6	7	8	9	10	11
Porifera			x	xx							
Cnidaria	x		x	x	xx	xx	x		xx		
Mollusca	xx	x			xx	x	x		xx	x	xx
Annelida			x								
Arthropoda											
Merostomata	x	xx	xx	x						x	
Arachnida			x					x		x	
Crustacea	xx	x			x			x			
Insecta					xx					x	xx
Onychophora										x	
Bryozoa					x						
Echinodermata	xx		x		x		x			x	x

xx, Major uses; x, sporadic or minor use.
[a]Main reasons: 1, human food; 2, animal food; 3, medical use; 4, household utility; 5, souvenir trade/ornaments; 6, jewellery; 7, aquarium trade; 8, pets; 9, industry; 10, research/teaching; 11, 'collectables'.

(as for Onychophora, below, *Limulus* is a 'living fossil' of considerable evolutionary significance), for animal fodder, fertilizer, eel bait, and for sale as tourist souvenirs. It is also killed directly by clam diggers to counter its depredations on clams.

Onychophora, velvetworms, have been regarded as possibly threatened in some areas by the demand for specimens from biological supply houses. Many Onychophora occur at extremely low densities and are very difficult to collect (even for experienced specialists) in large numbers. Others may be locally abundant in disturbed habitats or in caves (p. 24) and the latter, in particular, may be especially vulnerable. A few Onychophora species are known *only* from particular caves or cave systems, and intensive collecting there (where specimen 'return for effort' can be high and lead to rapid depletion of populations) can be endangering. Reliable figures for the numbers of specimens captured, and where they are caught, are understandably elusive for most invertebrates utilized in such ways.

The major reasons for possible overexploitation of various invertebrates are summarized in Table 3.5. For most, there are clear 'major' and 'minor' reasons, although the balance between these may vary widely within a higher taxon, and geographically. Most insects are not commercially desirable, for example, but a few groups are of intense commercial relevance, as described above. Trade for some species is prohibited or regulated (p. 114), and restrictions on amount or size of some species collected are numerous (p. 113). The collective human needs may impact on many different taxa in a habitat (Table 3.6).

Determination of the need for any such regulation, and of maximum exploitable levels to ensure sustainability of any species can come only from ecological study, and the high levels of numerical fluctuation which are normal in many invertebrate populations may necessitate long-term studies to determine them. Lack of this knowledge has led to cases of population endangerment in the interests of short-term gain. Rounsefell (1975) cited the beach-dwelling Pismo clam (*Tirela stultorum*) which was at one time so abundant in parts of California that farmers ploughed them out of sandy beaches at low tides, sometimes (up to the 1920s) using them as chicken food. The clam lives only on unprotected beaches swept by heavy surf, and was unable to withstand such intensive exploitation. It has recently been used only for 'tourist' fishery rather than on any commercial basis and has needed supervision to maintain it even at this level.

Not all exploitable invertebrate species are

Table 3.6 Major groups of commercially desirable marine invertebrates

Phylum	Items	Major uses
Porifera	Sponges	Industrial
Cnidaria	Precious corals	Jewellery, ornaments, tourism
Annelida	Bait worms	Commercial sale
	Palolo worms	Food
Arthropoda	Crustaceans	Food
	Horseshoe crabs	Food (processed), tourism
Mollusca	Mussels, oysters, other bivalves	Food
	Abalone, limpets	Food, bait
	Conchs	Food, shells for jewellery, tourism
	Cuttlefish, squids, octopus	Food
Echinodermata	Starfish	Industry, poultry food supplement
	Sea urchins	Gourmet food (roe)
	Sea cucumbers	Food (trepang)

removed permanently from their habitats. The stone crab, *Menippe mercenaria*, in Florida has very large claws and relatively little meat in the body. According to Rounsefell (1975), fishermen break off the claws and release the crabs to regenerate new ones. The effect of this mutilation on survival of crabs does not seem to have been established.

Especially in large-scale commercial fishing operations, catching of one species (or a group of species) can sometimes lead to harm to non-target invertebrates, either through habitat destruction or change caused by harvesting activities, or by being caught and killed through use of non-selective capture methods. This 'bycatch' may be substantial and diverse in some marine fisheries operations, and the effects are exemplified here by comments on the Australian prawn-trawl fishery.

Hutchings (1990) noted that about 1100 prawn trawlers are licensed to operate in Queensland coastal areas, which include the Great Barrier Reef. Collectively, these seek eight species of Penaeidae, with additional targets in some places including scallops (*Amusium balloti*) and scytharid lobsters (*Thenus* sp.). Three different trawling techniques are used:

(1) a net, towed between two trawlers is regarded as detrimental to epifaunal communities;

(2) a balloon trawl, which hardly touches the bottom; and

(3) a net towed at about 15 cm from the bottom, but with an advance 'tickler chain'.

As background to the faunal diversity involved, an appraisal of a series of trawls on the Great Barrier Reef yielded more than 700 species of sponges, coelenterates, molluscs, and echinoderms (Cannon *et al.* 1987). Hutchings noted also a series of trawls from the Gulf of Carpentaria: they included 251 species of molluscs, 145 coelenterates, 150 echinoderms, and 104 sponges.

Most bycatch from prawn fisheries is discarded overboard, so that discards from Moreton Bay alone are of the order of 3000 tonnes for a seasonal return of prawns of around 500 tonnes (Wassenberg and Hill 1990), with an average discard of 36 kg each passed net trawl and invertebrates constituting about three-quarters of this (52.2 per cent crustaceans, 17.9 per cent echinoderms, 3 per cent cephalopods). Only about 20 per cent of the animals are dead at the time of discarding, and the discard summed to around 1400 kg/night for Moreton Bay. Most of the bycatch invertebrates were thought to survive after discard, and the dead food was regarded as important to benthic scavengers.

In Torres Strait, prawn trawling is the most important commercial fishery (Harris and Poiner 1990; Hill and Wassenberg 1990) but invertebrates were relatively minor components of the bycatch, which consisted predominantly of teleost fish. Sporadic large catches of rock

lobsters were made. Fish were also the predominant bycatch in the Hawkesbury River (New
South Wales) (Gray *et al.* 1990), but 13 species
of Crustacea and five of cephalopod molluscs
were taken in this estuarine fishing ground, up
to 80 km upstream from the river mouth.

Survival of invertebrates depends on the
compaction of the catch in the trawl, and the
duration of exposure before discard. Cephalopods die rapidly when out of water: 10 minutes
exposure on deck after a 30 minute trawl time
is sufficient to result in death of cuttlefish
(*Sepia*) and squids (*Loligo*) within an hour
when replaced in water (Hill and Wassenberg
1990). By contrast, 88 per cent of starfish ($n =$
'at least 30') and all portunid crabs were alive
after 12 hours.

Globally, discards from the prawn-trawling
industry amounted to around 1.4 million tonnes/
year in the early 1980s (Saila 1983). Assessing
the effect of this excess take on epibenthic
communities is difficult, because of very sparse
'before and after' baseline data. Likewise, effects
on related infaunal communities are unclear.

The threats discussed in this chapter are all
proximal, in that they are occurring and must
be countered to the best of our ability. Other
threats are less tangible: overriding all of those
noted here, for example, are the ramifications
of global climatic change, a topic discussed on
p. 159.

FURTHER READING

Di Castri, F., Hansen, A.J. and Debussche, M.
(ed.) (1990). *Biological invasions in Europe and
the Mediterranean Basin*. Kluwer, Dordrecht.

Drake, J.A., *et al.* (ed.) (1989). *Biological invasions: a global perpsective*. Wiley, New York.

Hart, C.W. and Fuller, S.L.H. (ed.) (1974). *Pollution ecology of freshwater invertebrates*. Academic Press, New York.

Heliövaara, K. and Vaisanänen, R. (1993). *Insects
and pollution*. CRC Press, Boca Raton, Florida.

Hellawell, J.M. (1986). *Biological indicators of
freshwater pollution and environmental management*. Elsevier, London.

Jepson, P.C. (ed.) (1989). *Pesticides and non-target
invertebrates*. Intercept, Wimborne, Dorset.

National Academy of Sciences (1975). *Petroleum
in the marine environment*. National Academy
Press, Washington, DC.

Nichols, D. (ed.) (1979). *Monitoring the marine
environment*. Institute of Biology, London.

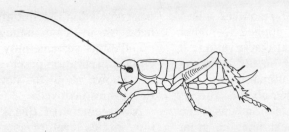

4   PRACTICAL INVERTEBRATE CONSERVATION: APPROACHES AND SETTING PRIORITIES

FORMULATING APPROACHES

The great diversity of threats to invertebrates discussed in the last chapter necessitates correspondingly diverse responses in attempts to conserve the animals. Two major approaches have played important roles in this, and these reflect different levels of focus:

(1)   the 'species level', where particular species or other taxa have been the primary conservation target; and

(2)   the 'habitat level', where representative habitats have been the primary target for reservation or protection, with the assumption that the taxa present will thereby be conserved effectively.

Both these approaches have merits and shortcomings, and pose questions of 'what to conserve' and 'how to set priorities'—questions which are extremely difficult to answer but necessary in practical conservation planning where logistic capability is insufficient to meet all demands. The grounds for setting priorities, whereby some invertebrate taxa or communities are ranked ahead of others and the latter in essence, condemned, raise intricate ethical questions, for example. Yet such rankings are continually being made, and this practice seems set to continue. Many different parameters for priority have been incorporated in this broad process, which is commonly underlain by some assumption of

'triage'—that the resources available can be used to assure the greatest 'good'. Broader approaches to 'biodiversity conservation' are currently being developed but the two major levels of practical focus until now are discussed here.

They differ substantially in emphasis. At the species level, 'taxon values' are paramount, so that vast amounts of effort may be spent to ensure (or attempt to ensure) the conservation of particular notable species, or even subspecies, when these are endangered or perceived to be threatened in some way. At the habitat level, the emphasis moves to entire natural systems or communities, often reflected by particular sites, and away from particular species. Indeed, managing a site for a given species may well render it unsuitable to other taxa. One universal problem with attempting to manage single species or groups is that the repercussions on other species are likely to be almost entirely unknown. Together, the two approaches reflect the different levels at which conservation activities are perceived, and their various advantages and shortcomings are noted here. They can, of course, operate together, because a high-priority habitat is commonly one which harbours priority species, and habitat protection is the major need for any vulnerable species or assemblage. Briefly, many authorities claim that 'species-focusing' *per se* in the face of the enormous diversity of invertebrates which may demand such intensive attention because of their vulnerability, is a 'waste' of resources which could be of more benefit if applied to higher levels of community or faunal conservation. It has been likened to focusing on the tip of an iceberg (Fig. 4.1), in paying attention simply to the few taxa seen to be vulnerable or threatened and which have by some means been accorded priority. In the meantime the great hidden majority of taxa may be lost progressively and, unless the focal taxa are functionally important, the communities or systems of which they are part may become unstable anyway. One major drawback of species-focusing is the vast expense involved in prosecuting the sequence shown in Fig. 4.2. The 10 top-ranked species listed under the United States Endangered Species Act, none of which is an invertebrate, took up about half the total funding available in 1990, for example, and 58 species (fewer than 10 per cent of the 591 listed taxa) accounted for 90 per cent of the expenditure (LaRoe 1993). It is clear that without massively increased funding for species conservation programmes, no more than a tiny proportion of needy taxa can ever be addressed effectively—even if they are high-profile mammals or birds.

The counter-view is that habitat conservation may not, in any case, afford protection to all the significant taxa present, despite them being assured of a place to live, and that these must be given specific attention to assure their well-being. Without a habitat, of course, a species cannot survive in the wild, but the requirements of many invertebrates are so specialized that they might not survive anyway in a habitat which is merely 'locked up' and not managed in any way. Many of those taxa characteristic of advanced successional vegetation types or, more generally, of mature ecosystems, are likely to do so, but many that depend on early or intermediate successional stages, or on less mature ecosystems, may well disappear as the habitat changes naturally over time, unless it is managed in some way to assure the continued availability of such habitats. This principle has been demonstrated repeatedly with the British butterflies, for example (Thomas 1984), where many populations have become extinct on nature reserves, including some reserves

designated for the butterflies themselves. The need for intricate management to assure the populations' sustainability was not appreciated in early conservation practice, because insufficient was known about the detailed requirements of the species involved.

Management of this kind, so necessary at the 'species level' of conservation, depends on detailed knowledge of the ecology of the taxon involved. More generally, much conservation management also necessitates countering threatening processes in other ways, such as by regulation (p. 112), rather than just buffering the habitat from those threats. The sequence of steps involved in practical conservation targeting a given species is summarized in Fig. 4.2. There are several major steps in this general sequence.

1. Establishing the status of the species gives a strong foundation for all subsequent management.

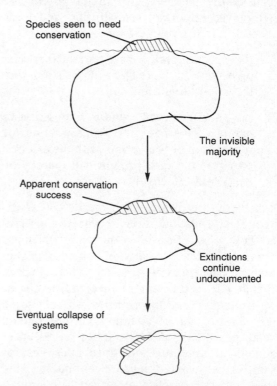

Fig. 4.1 Species-focusing as 'the tip of the iceberg': unless the most significant species are selected, the majority may still not be sustained by this approach, and ecosystems or communities may eventually collapse.

Very broadly, the components of status include the taxonomic and biological integrity of the species, its rarity, and how its abundance and distribution may be changing. In many instances a species is recognizable easily as a distinctive entity, but this is not always so. Many butterflies, for example, have been given trinomial (subspecific) names for local populations of dubious biological distinctiveness, but these have attracted considerable conservationist attention. Especially in the northern hemisphere, much attention has been paid to localized or threatened subspecies rather than to full species; because many have small geographical ranges, they can be vulnerable to localized threats which might not influence the parent species as a whole. In some protective legislation (p. 114), a subspecies is deemed equivalent to a full species if it is readily accepted by scientific consensus (Opler 1991). Field surveys may lead to a taxon being adjudged safe or not. If the latter, it may be 'vulnerable', in which case monitoring to detect future changes is highly desirable, or 'threatened' because decline is occurring.

2. Establishing the nature of threats and the time scales over which they operate is needed to counter them. Causes of decline in range or abundance may be natural (as by succession in habitats) or anthropogenic, and the latter can operate over different time frames. We may be seeing a residual effect due to threats now past, current threats in operation over a finite period, or permanent change with effects likely to persist far into the future.

3. Determining or anticipating the effects of these on the organisms themselves or their environment leads to the next step.

4. Defining the kind of management strategy needed, probably with an action plan (or similar) formalizing the steps in a logical sequence over a given period.

5. Implementation of management is then needed to achieve one or more of threat abatement, habitat protection and species conservation. The latter has two rather different contexts. It may be adequate to prevent decline from current population levels or, if substantial decline has already led to small, possibly nonviable populations, a more aggressive strategy aimed at 'recovery' may be needed.

6. Any management should be monitored carefully, leading to continually refining details or approaches as the programme proceeds. Any conservation management is dynamic and must respond to changes in species or habitat condition or vulnerability. Monitoring is thus usually site-based.

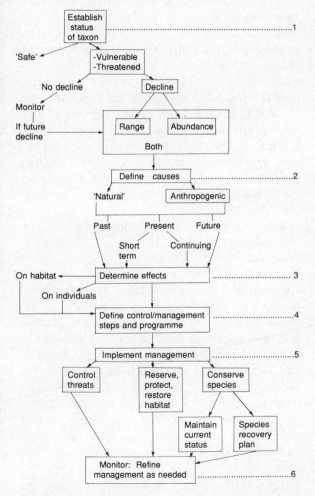

Fig. 4.2 A sequence of steps for a single species conservation plan.

However, lack of ecological knowledge for all except a minute proportion of invertebrates means that the *only* viable step for most taxa is to concentrate on the habitat level. No other option is possible for diverse faunas, and the problem then centres on selection of optimal habitats or sites to be reserved, when a choice is available. Often, there is no choice, but one consequence of high diversity of invertebrates is that any given site is probably unique in details of its invertebrate complement or, at the least, is difficult to cross-match fully with any other site.

The need to reserve examples of as many different habitats as possible is recognized widely, but the ranking of possible reserve sites within any such habitat to ensure that the 'best' one(s) is chosen is extraordinarily difficult, because of the wide array of values that need to be considered, commonly in a short time. In addition, the great variety of geographical scales and habitat subdivisions 'perceived' by invertebrates need to be considered in selecting optimal areas for conservation.

A number of criteria relevant to priority taxa of invertebrates have been discussed (Usher 1986), but the presence of 'notable' taxa (p. 76) can prove significant, so that the two approaches delimited earlier can be complementary in effecting practical conservation measures.

Protocols are gradually being developed for both habitat and species approaches and, because of the great variety of detail needed to cope with a sufficient range of cases, these are perhaps approached most constructively through a 'check list' of points which need to be considered. In general, habitat reservation or protection is a prelude to management, but in the absence of defined, or definable, management needs, it is in itself essential.

Thus, for 'habitats' to conserve representative diversity, Kirby (1992) emphasized the need to understand the use of the habitat by invertebrates and how this might be changed by past, current, and anticipated threats, as the background to determining the importance of habitat subunits. Major needs at a site are continuity of habitat and adequate structural variation. Thus for woodland in Britain, the major components

were trees and shrubs, dead wood, and open space with understoreys and margins. As with most other natural habitats, such a complex system provides many 'subhabitats' for different invertebrates, and all are needed to sustain the natural diversity present. The management practices exemplified in this chapter are thereby parts of a complex suite of tools to retain these— from the range of plants and different growth stages, to accepting the value of dead wood (including stumps, fallen trunks and branches, dead standing trees, tree holes, and others) and the various light/shade regions and understoreys of open ground. Many such conservation measures are founded in traditional practices such as coppicing, and conservation need has come only since these were abandoned, and habitats reduced to small patches of their former distribution.

More generally, management needs may include a range of successional stages and structural variation, and Kirby (1992) emphasized also the importance of rotational management to maintain these. This means, simply, that only a fraction of a site is changed (mown, grazed, logged, or other) at any one time so that vegetation is maintained in the greatest possible suite of ages. More abrupt or large-scale change could constitute a major threat to invertebrates.

Management for species, following the general scheme depicted in Fig. 4.2, integrates many of the topics discussed separately in this book, in essence combining these into a coherent and dynamic system. Arnold's (1983) pro-forma scheme for species-orientated conservation targetting rare lycaenid butterflies (Table 4.1) indicates the parameters and complexity needed. Many lycaenids (the blues, coppers, and related butterflies) participate in complex ecological interactions in that their caterpillars have specific foodplants but are also tended by particular ant species and cannot persist without the ants which, in turn, may protect the caterpillars from predators and parasites.

Thus their management, must provide for incorporating such vital and intricate associations (New 1993a). Arnold's scheme illustrates the

Table 4.1 A pro-forma scheme for species-orientated conservation of invertebrates: a scheme designed for rare lycaenid butterflies (Arnold 1983)

1. Preserve, protect, and manage known existing habitat to provide conditions needed by the species.
    (a) Preserve; prevent further degradation, development, or modification
         (i)   co-operative agreements with landowners and/or managers
         (ii)   memoranda or undertakings
         (iii) conservation easements
         (iv)   site acquisition (purchase/donation of private land) or reservation (public land)
    (b) Maintain land and adult resources
         Minimize threats and external influences
    (c) Propose critical habitat
    (d) If recovery, clarify taxonomic status of taxon in habitat and other populations
2. Manage and enhance population(s) by habitat maintenance and quality improvement, and reducing effects
    of limiting factors
    (a) Investigate and initiate habitat improvement
         methods as appropriate
    (b) Determine physical and climatic regimes/factors needed by species
         and relate to overall habitat enhancement at site
    (c) Investigate ecology of species
         (i)   life style and phenology; dependence on particular plant species or stages
         (ii)   dependence on other animals, and their roles
         (iii) population status
         (iv)   adult behaviour
         (v)   determine natural enemies and other factors causing mortality or limiting population growth
         (vi)   investigate possibility of captive breeding for introduction or translocation
    (d) Investigate ecology of tending ant species, if present
    (e) Investigate ecology of foodplant species
3. Evaluate all the above and incorporate into development of long-term management plan. Computer modelling may
    assist in making management decisions
4. Monitor population(s) to determine status and evaluate success of management
    (a) Determine site(s) to be surveyed, if choice available
    (b) Develop methods to estimate population numbers, distribution, and trends in abundance
5. Throughout all of the above, increase public awareness of the species by education/information programmes
    (such as information signs, interpretative tours, audio and visual programmes, media interviews, etc.)
6. Enforce available regulations and laws to protect species. Determine whether any additional legal steps are
    needed, and promote these as necessary

difficulty of formulating any general 'recipe' and one purpose of this chapter is to indicate the kinds of information needed to refine a protocol for any given species' management. Particulars of land (site) tenure, external threats, species biology, and so on will vary from case to case. It is important to incorporate all these political, legal, biological, and monitoring aspects into a viable programme for effective management. Similarly, the details of methods for population estimation will differ in individual cases, and the resources available. In general, though, a combination of methods may be needed— for the lycaenids, it may be useful to undertake reasonably independent counts of adult

butterflies and of caterpillars on their foodplants, for example.

PRIORITY TAXA: THE 'SPECIES APPROACH'

Any of the values for invertebrates noted in Chapter 2 can be used to promote the worth of particular taxa as foci of conservation need, and the various parameters of ecological worth are vital in suggesting priority 'groups'. Public appeal or 'charisma' can be very important in fostering sympathy, and threats to species with tangible commodity values are also likely to receive support from those who benefit from those values! However, a number of the specific cases noted in Chapter 7 do not derive clearly from any such value, and less tangible parameters which can give precedence for conservation include scientific status and importance (such as taxonomic isolation), 'rarity', and the perception of an imminent threat to any species which gains a vociferous advocate. The latter syndrome has resulted in a number of taxa, suffering especially from local threats in industrialized countries, gaining massive, but largely uncoordinated, support as a result of individual zeal, which would be difficult to defend on more pragmatic grounds. Indeed, in wealthier parts of the world the condition of species 'worthiness' for conservation can depend simply on 'whether or not we like it', if more instrumental values are not obvious (Regan 1986), and plays an important·role in fostering awareness of conservation need. Reliance on instrumental values for setting priorities leads to what some cynical authors refer to as the 'use it or lose it' aspect of conservation priority! The ethics involved in conservation, of any group of organisms, are immensely complex.

'Species desirability' (Adamus and Clough 1978) thus includes a number of parameters, including those described below. Scientific status and rarity have been the most frequent criteria used to select invertebrates as conservation targets. This is also the case for vertebrates, for which the option of species-focusing has been the major avenue to conservation progress.

Taxonomic isolation

The degree of taxonomic isolation reflects the genetic and biological peculiarity of the taxon concerned, simply whether it has close relatives. Species that are the sole extant representatives of distinctive genera, or even of families or superfamilies, are ranked more highly than species with many congeners because their loss would represent extinction of a higher taxon or a more isolated lineage. Such isolated taxa are often of ancient lineages and are relicts

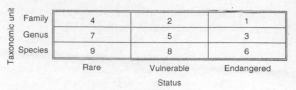

Fig. 4.3 Taxonomic priorities for allocating conservation effort (after Xu Zaifu 1987). IUCN status categories increasing in intensity from rare to endangered; higher taxonomic units take priority over lower ones; numbers denote priority sequence from 1 to 9.

or 'living fossils' (Ghiselin 1984), and many of them have unusually high evolutionary significance. Examples are the species of *Nautilus* and the unusual *Neopilina* among the molluscs, and the relict damselfly *Hemiphlebia mirabilis* (p. 135) and the two extant species of the dragonfly suborder Anisozygoptera (one in the Himalaya, one in Japan) in the insect order Odonata. A rationale for a taxonomic sequence priority scheme (after Xu Zaifu 1987) is shown in Fig. 4.3. In this, the sole representatives of categories higher than family would have a persuasive case for conservation if they were seen to need active intervention to ensure their well-being. Many ancient invertebrate forms, of the kinds exemplified above, are indeed rare and some are known to have declined in recent decades.

Rarity

Both for higher groups and for species within each group, 'rarity' can be a persuasive feature for demonstrating conservation need: it is equated by many people to 'in need of conservation' but, of course, the proper concern relates more to perceived decline in abundance and the factors or processes engendering this, because vast numbers of species are naturally rare. Rarity, and decline, can be defined in several different ways, based on numbers, distribution, or a combination of these. Thus, as examples, a species might be adjudged rare because it occurs in low numbers over a wide range, or because it occurs in only one (or few) places even though the population(s) there might be reasonably large. This situation is sometimes referred to

Table 4.2 Dimensions of species rarity (after Rabinowitz 1981; Cody 1986), for explanation see text.

| Combination | Category of rarity | | | Status |
	Alpha	Beta	Gamma	
1	Common	Common	Common	Common
2	Common	Common	Rare	
3	Rare	Common	Common	Rare: level 1
4	Common	Rare	Common	
5	Common	Rare	Rare	
6	Rare	Common	Rare	Rare: level 2
7	Rare	Rare	Common	
8	Rare	Rare	Rare	Very rare

In each category, a species may be adjudged 'common' or 'rare', with the result shown.

as 'local'. The most significant category of rarity is of a single population with low numbers, and conservation concern is highest if:

(1) this sole population has been documented as declining in numbers, or/and

(2) other populations have been known in historical time but are now extinct, or

(3) the population is perceived to be threatened in some way.

Rabinowitz *et al*. (1986), Cody (1986), and others discern seven categories of rarity, depending on the abundance of a species within a community (if scarce, it displays 'alpha rarity'), its distribution along a gradient of habitats (reflecting the degree of ecological specialization; extreme specialists have high 'beta rarity'), and its geographical range (if this is small so that the species is narrowly endemic, it has high 'gamma rarity'). The various combinations of these states (Table 4.2) show different ways in which a species can be rare, with the extreme case of low abundance, narrow habitat tolerances, and restricted range. Even then the situation might be stable and sustainable. Such species might easily be made vulnerable, but are not automatically priorities for conservation action.

Evidence of decline or threat to such a population may be conspicuous but, conversely, decline may be difficult to differentiate from normal population fluctuations, which can result in populations being naturally at low levels over several generations. Even extinction may not be easy to confirm for some populations or species, for reasons allied to habits and population structure. As examples of such difficulty:

1. Some naturally sparse invertebrates can be very hard to find because of their cryptic habits. For Onychophora, for example, Mesibov (unpublished manuscript, 1991) commented that 'velvetworms are hard to find not so much because they are rare, but because they're truly expert at hiding'. Such invertebrates, for which quantitative searching effort by specialists can yield very few specimens/hour of effort (Scott and Rowell (1991) on the Australian velvetworm *Euperipatoides leuchartii*), may often appear to be rare (or extinct) whereas this is a relative impression based on the species' biology.

2. Many invertebrates have a structure of 'metapopulations', by which (rather than a single integral population) *groups* of predominantly self-contained local populations within a habitat or site maintain the potential for dispersal—albeit sometimes infrequently—between them. Extinction of a particular local segregate may lead, in time, to colonization of that site from other population segregates in a pattern of rolling extinction–restoration cycles, as an integral facet of population structure. Exchange of only a few individuals, comprising a very low proportion of total numbers, between local 'colonies'

may buffer against numerical fluctuations and spread the risk of extinction (Den Boer 1968), and an important facet of habitat management for such taxa is to ensure that barriers to such dispersal are not imposed inadvertently. Differences in dispersal powers occur even between similar and closely related species, so that it is difficult to generalize in predicting what constitutes 'isolation' for any given taxon until it has been studied directly.

However, a 'typical metapopulation structure' poses other problems for appraisal, as exemplified by Ehrlich and Murphy's (1987) study of checkerspot butterflies, *Euphydryas* spp., in California. In their most intensively studied case, of Edith's checkerspot (*E. editha*) on Jasper Ridge, one of three main demographic units monitored since 1960 went extinct in 1964, re-established in 1966, and again became extinct in 1974. Potential for dispersal of butterflies increased in certain dry years when oviposition plants and nectar sources for adults were sparse. Density-dependent dispersal occurred very rarely, but these years may be the main ones in which recolonization of marginal or small habitat patches occurred. Recolonization of empty patches may not occur in most years and the amount of butterfly dispersal varied greatly in relation to precipitation and the extent of post-diapause larval defoliation of foodplants before the main oviposition period. Populations can persist at very low densities if other factors remain favourable. Some parts of the overall population are *always* extinct, with their sites awaiting recolonization, so that the habitat patches are vacant. Under the US Endangered Species Act (p. 116), these habitat patches then have no legal protection (Ehrlich and Murphy 1987). Conservation of metapopulations may be difficult, as alienation of some habitat units would interfere with dispersal and recolonization to the extent that the multiple system would become extinct. The management possibilities for metapopulations of *E. editha* were examined further (Murphy *et al.* 1990), and there is a clear need to understand the metapopulation dynamics in order to interpret the butterfly's regional distribution at any given time.

Thus, even that a species (population) is not seen at all at a given locality for several generations or years may not mean, necessarily, that it has become extinct.

Determining the conservation status of rare species, and formulating a recovery or management plan (p. 73) is a frequent need in invertebrate conservation. Often, the needs appear to be simple. Protection of the site(s) is paramount, but understanding the species' ecological needs and likely response to threats is also necessary. However, it is usually not possible to 'experiment' with such rare or vulnerable species, and even attempts to estimate changes in population size from generation to generation may be severely restricted. Likewise, it may not be possible to plan suitable controls for management (Usher and Jefferson 1991) because the species may be present at only a single site, and then only in small numbers. Attempts to use such techniques as mark–release–recapture may not be appropriate: this technique is used frequently for butterfly population estimation, for example, but might not be suitable for small, delicate taxa because of possible adverse handling effects harming the insects (Murphy 1989). Even straightforward observation might be difficult because of the low numbers present, and any active searching might constitute disturbance to the population or habitat, such as by trampling low-growing foodplants. Observations might, thus, be restricted to the most conspicuous life-stage over a short period of the year, at a single, small site which harbours the only known breeding colony or population of the taxon. Broader sampling techniques (p. 129) which normally trap and kill invertebrates are not suitable for such rare species, for which any untoward mortality must be avoided so that populations are not rendered even more vulnerable.

In short, rare species are difficult to study, and establishing their status is costly.

For conservation of rare species, Hooper (1971) noted four possible management options, which remain valid decades later:

(1) they are not conserved directly;

(2) a single large reserve with large populations is protected;

Table 4.3 Principle of critical fauna analysis: data on swallowtail butterflies (Lepidoptera: Papilionidae) (from Collins and Morris 1985)

		Number of swallowtail species			
		Endemic	Total	Cumulative total	% of family represented ($n = 573$)
1.	Indonesia	53	121	121	21.1
2.	Philippines	21	49	146	
3.	China	15	104	222	38.7
4.	Brazil	11	74	296	
5.	Madagascar	10	13	309	53.9
6.	India	6	77	323	
7.	Mexico	5	52	365	63.7
8.	Taiwan	5	32	370	
9.	Malaysia	4	56	375	
10.	Papua New Guinea	4	37	387	67.5

Countries are listed in descending order of number of endemic species, and if two countries have the same number of endemic species, they are listed in order of the greatest number of other species, which are added to the collective total. The top 10 countries are shown.

(3) several sites within the normal dispersal range need to be conserved; and

(4) species are continually reintroduced from captive populations (Chapter 5), needed especially when the site has very restricted carrying capacity.

Critical faunas

An extension of the 'species-focus' rationale is the detection and ordering of so-called 'critical faunas', where aspects of endemism and diversity of a given taxonomic group are combined so that a number of rare or local species may have a combined impact on determining conservation, rather than, for example, a site being selected on the basis of only one species of note. This approach has been expressed recently for two of the best-known groups of butterflies, the milkweeds (Nymphalidae: Danainae; Ackery and Vane-Wright 1984) and the swallowtails (Papilionidae; Collins and Morris 1985). This kind of appraisal is appropriate for groups whose distribution and systematics are well documented, or for which such documentation can be obtained relatively easily.

The scale over which endemism is determined

can be varied, either on a country, smaller political unit, or geographical unit or site basis, so the method is of value for ranking sites as well. The swallowtails were assessed on a country basis but the geographical complexity of the milkweed butterflies in some diverse archipelagos such as Indonesia and the Philippines, necessitated their assessment on various subdivisions of these countries, for example. The principles of the method are exemplified in Table 4.3.

Countries (or other areas) are ranked downward from those with the highest number of endemic species, in a sequence of maximum complementarity. Thus, from Table 4.3, Indonesia has the greatest number (53) of endemic swallowtails. The second highest number (21) is for the Philippines, and adding the *total* species for the two areas covers 146 species. Adding in sequence shows that, if the 'top five' countries' faunas could be conserved, more than half (*c.* 54 per cent) of the global representation of the family would be present. The top 10 countries in total increase this proportion to around 68 per cent, and 43 countries had endemic swallowtails.

The principle involves ensuring complementarity, so that if two areas have the same number of endemic taxa, the higher-ranked would be that which adds the greater total number of 'new' species to the previous listing. The aim is to define the minimum suite of areas needed to support at least one population of every species in the group concerned. Areas are added progressively until this is achieved. Clearly, all areas with endemic species must be included, by definition, and it may be necessary to add others to assure complete representation. For the Danainae this process defined 31 areas to include all 157 species; for the Papilionidae, 51 countries were needed to represent all 573 species.

Vane-Wright *et al*. (1991; see also Williams *et al*. 1991) have recently extended this principle of selecting 'critical faunas' to nominate areas with high 'biodiversity value', incorporating consideration of taxonomic rankings, and the process is likely to prove important in helping to set priorities for many different taxa. In terms of site reservation, cost-effective conservation requires the maximum biodiversity to be protected in the minimum area (Williams *et al*. 1991). Once an area is protected, the next most valuable is that which adds the greatest representation of additional taxa, and so on.

HABITAT PROTECTION: RESERVES

Protection of natural habitats, together with their range of specialized biota, from the range of threats and intrusions noted earlier is the main thrust and need of invertebrates. Without information on the detailed needs of the vast majority of invertebrate taxa (a situation which will not improve significantly in the foreseeable future), the only practical option for conserving the bulk of natural biodiversity is to protect the greatest possible range of habitat types, and number of sites within each of these, in the expectation (and hope) that many of the occupants will thereby also be protected.

Habitats, sometimes represented by particular sites, are the prime vehicle for ecosystem and community conservation, but emphasis on this level brings practical problems not apparent when considering species alone. We are dealing with complex systems, usually without sharp or easily defined boundaries, which intergrade in various ways. While a pond or stream may be defined as a habitat, assuring that it remains in suitable condition may necessitate managing adjacent areas, even as large as whole drainage basins or catchments. Political aspects include having a range of property owners or managers. In short, management for 'diversity' can also be expensive, and requires a holistic approach.

Not only unique or unusual areas need to be protected. There is an increasing need for adequate representation in reserves of what may be considered at present to be 'typical' and widespread habitats, such as grasslands and others, which are being subjected to increasing and insidious change and invasions by exotic species. Indeed, total representativeness—keeping all the species—is the primary goal for an optimal system of nature reserves (Vane-Wright 1994).

Disney (1986a) recognized three main strategies for nature reserves:

(1) saving endangered species considered worthy of conservation. Site identification depends on detailed ecological knowledge of the species involved;

(2) establishment of an adequate series of protected nature reserves, including 'key' sites; and

(3) re-establishment and maintenance of a landscape including a high density of wildlife refuges.

His striking analogy (Disney 1975) that 'scheduled reserves can be regarded as the spectacular motorway system by which we hope to convey a diverse wildlife into the future' implies the range and complexity needed.

These levels reflect species protection, community protection, and maintenance and restoration—or, even, creation of new habitats where no similar ones had existed before. Webb (1989) gave a slightly different set of reasons, but with a similar joint emphasis:

(1) conservation of populations of rare or endangered species;

(2) conservation of areas where species richness is high; and

(3) conservation of assemblages of species which represent the natural or seminatural communities characteristic of a region.

The term 'critical habitat' is used commonly in discussing the reservation needs for particular taxa. It is defined by the United States Endangered Species Act as not only the geographic area where endangered and threatened species occur at the time they are listed, but also 'specific areas outside the geographic area occupied by the species at the time it is listed . . . upon determination by the secretary that such areas are essential for the conservation of the species'. The concept can thereby incorporate the less obvious needs for species, but critical habitat can be very hard to delimit adequately. It can, for example, incorporate refuge areas used for hibernation or aestivation in addition to the breeding habitat; for some butterflies hill-topping or assembly sites used for courtship and mating may be some way from these. Topographic prominences, not necessarily large, are critical to the well-being of many butterflies.

Because of the immense diversity of invertebrates, it is unlikely that the fullest practicable reserve system would afford complete representation of all communities, as each site is likely to harbour a unique suite of taxa (New 1984). However, although conspicuous vertebrates and vascular plants, particularly, have served as the predominant foci (and 'umbrella species') for reserve selection, there is no doubt that the information available on ecosystem dynamics from the invertebrates present is substantial, and that they could play far greater roles in (1) assessment of sites for reservation and (2) ranking of sites, such as those based on a particular vegetation type or coastal landform, when it is not possible to protect them all. Consideration of assemblages of invertebrates, including diversity and presence of critical faunas (p. 79), exemplify the emerging awareness of this.

For the habitats of many rare species, the conservation 'choice' is simple: protect the site on which it occurs, or condemn the taxon to extinction in the wild. This is the usual predominant thrust of species-orientated conservation. The first category of reserve defined by Disney

(1986a) or Webb (1989), therefore, may pose little conceptual problem. Practical problems can, though, be substantial as the need is often to purchase, or otherwise secure, a critical habitat in the face of opposing interests.

Most instances are not as clear-cut, and the subtle differences between invertebrate assemblages may lead to many unheralded demises of species associated even with adequately represented habitats; ideal levels of protection for all species in any community are not realistic to achieve, and the task is to 'do the best we can' on criteria as objective and biologically realistic as possible. It will, I suspect, always be easier to secure a terrestrial reserve for a rare mammal or bird than for an insect, spider, snail, or worm— although examples of all the latter do indeed exist—but reserves under such politically powerful umbrella species as charismatic vertebrates may represent the best that can be achieved for invertebrate habitat security in some places.

It has been emphasized frequently that reserves for invertebrates need not, necessarily, be large. They can, conversely, be too small, and one general principle of reserve design (below) emphasizes the desirability of larger reserves over smaller reserves. In some instances, assuring the integrity of a reserve may necessitate protecting a large area—the total catchment of a stream, for instance. However, in many places, we are committed already to a limited choice of small sites or areas—small patches of remnant vegetation in highly altered landscapes, of unpolluted wetlands, or short lengths of marine coastline not yet subjected to degradation. Marine reserves are indeed important for invertebrates, not least because the general continuity of the marine habitat (with many taxa having a pelagic planktonic life-history stage) enables continued existence of breeding populations, which may then recolonize devastated areas; Johannes (1978) gave an appraisal of traditional marine 'reserves', some of which have existed for centuries (p. 125). In Oceania, practical steps to ensure sustainability of marine resources, and derive the direct benefit from these, included systems of reef and lagoon 'tenure' by clans or family units. In such cases, it is likely that the 'commodity species' may have served as an umbrella for less conspicuous invertebrates in those areas.

Gaining protection of sufficient sites and habitats to ensure that an adequate suite of invertebrate assemblages is included as 'representative biodiversity' is difficult in all major ecosystems because of conflicting demands for land and water use. Barnes (1991) noted the case of British coastal lagoons, many of which are subject to change (either by succession to freshwater lakes and reed swamps, or by loss through erosion or changes of enclosing barriers), but which contain species and assemblages not found in nearby aquatic systems except those made by people during land reclamation or coast protection schemes. 'Protection' of such assemblages is therefore largely fortuitous, with the lagoons not being the primary object of management *even if* they are within nature reserves. This situation is not uncommon and highlights a need to assess priorities at a more subtle level than merely 'locking up species' in reserves.

In general, though, reserve areas created specifically for invertebrate species or assemblages are rare in relation to other reserve categories.

Reserve design

The purpose of a reserve can determine its design, and limitations are likely to be dictated also by the availability of particular sites in relation to other constraints and needs. In general, because populations of many of the invertebrates that have been the subject of species-focus conservation efforts are small and isolated (hence, their vulnerability and the concern for their well-being), only very small areas may be needed or available to contain viable populations. Unlike requirements for many vertebrates, a very few hectares may be sufficient to maintain particular focal species and representative invertebrate communities. On the other hand, reserves to maintain 'biodiversity' should, ideally, be large, and the optimal design of nature reserves has received considerable attention. Many of the principles are derived directly from considerations of island biogeography (MacArthur and Wilson 1967; Diamond

1975), from the analogy that a potential reserved area is a natural island in a 'sea' of disturbed or hostile habitat and may thus show substantial parallels with oceanic islands. Even small habitat islands are of considerable practical importance in invertebrate conservation. This means, for example, that effective reserves can sometimes be made within the context of large-scale and intensive land modification for agriculture, as in much of Europe. The small remnant patches of natural vegetation in the Western Australian wheat-belt are another example of this.

In some instances, remnant habitats include corridors (such as hedgerows, roadside verges, or other narrow strips such as streamside vegetation) which are the sole or preferred habitats of taxa that have been eliminated from adjacent areas through change, and thus they assume significant conservation value in their own right. Remnant habitat patches function as 'refugia' for a great variety of invertebrate taxa; Key (1978), for example, noted the dependence of some flightless endemic Australian grasshoppers on early pioneer cemeteries in country areas, often the only areas enclosed against the effects of grazing stock and thus harbouring natural vegetation. In other situations, a different kind of 'corridor'—such as cleared strips, woodland rides, or roads—can constitute formidable barriers to invertebrates (p. 90) and fragment habitats to the detriment of populations.

However, isolation can serve to impose barriers inadvertently between populations, and knowledge of the population structure and dynamics of a given species must be evaluated in trying to predict the usefulness of any small reserve erected to conserve that species. Many invertebrates, perhaps more than we suspect, form closed populations in which there is naturally little or no migration and interchange of individuals with other breeding units. Further, the kinds of habitat surrounding any given reserve may influence its integrity substantially, by either isolating it, or by complementing its biota in various ways (see Fig. 4.4).

General caveats on reserve design therefore relate to (1) size of reserves and (2) their degree of isolation, in relation to vulnerability of any target taxa (Fig. 4.5). Reserves can, at the

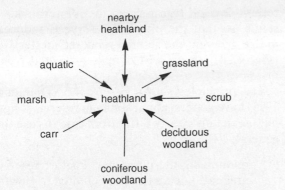

Fig. 4.4 Isolation of habitat. Example of interactions with surrounding communities which may influence integrity and species composition: 22 heathlands in Dorset, southern England. Where structurally more diverse vegetation surrounds a patch of heathland, there is a tendency for greater richness of invertebrates on that heathland. Only with grassland was there no tendency for this to occur. (After Webb 1989.)

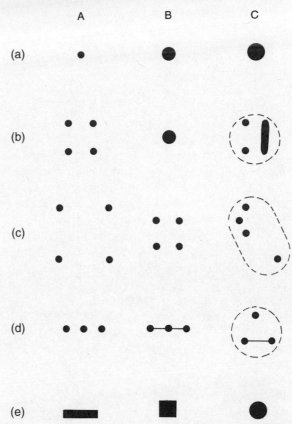

Fig. 4.5 Principles of reserve design (based on Diamond 1975). Conditions A, B, and C represent successively preferable states of parameters (a)–(e), respectively: (a) increase in size; (b) single large or several small (see text), minimum size can be passed and very small reserves may not be viable; if several reserve patches, can they be enclosed in a buffer zone (dashed line); (c) closer reserves better than widely spaced ones, to facilitate dispersal between them; (d) if possible, provide a corridor for linear reserve patches, or avoid a linear arrangement; (e) for reserves of similar area, minimize edge : area ratio.

extreme, be too small to protect a sustainable population or community, and too isolated to permit normal interchange of individuals between population segregates in taxa where this is a normal part of their dynamics. Reserve shape is also relevant, especially for small areas, because of the need to buffer against changes induced by peripheral disturbances, and it is highly desirable that an intermediate 'buffer zone' is established between the reserve and a surrounding highly disturbed region to help counter these effects or to 'soften' their impact, perhaps by imposing a different management regime there. As emphasized earlier, reserve design should, wherever possible, go hand-in-hand with projected management and, if the site is characterized by particular significant assemblages, their needs should be considered.

For woodlands, for example, this could be an area of low-intensity forestry between forest plantations and reserves; for agriculture it may take the form of unsprayed headlands (p. 93); in urban areas, it may be a region of land free of buildings and concrete, and so on. An effective buffer zone, therefore, is not necessarily subject to the same form of protection or management as the reserve it protects.

Many invertebrate communities demonstrate the 'edge effect' (Fig. 4.6) markedly, and such buffers serve to dilute and impede the penetration of external impacts to the reserve proper. They may also be areas where such threats can be controlled with little risk of harm to the main reserve. Enlargement of the reserve, to the extent that the 'core' is effectively protected (Fig. 4.6b) is highly desirable; creating a larger intermediate buffer (Fig. 4.6c) may dilute further the impact of threatening processes, and without a buffer zone (Fig. 4.6a) much of a small reserve may be disturbed. This scenario

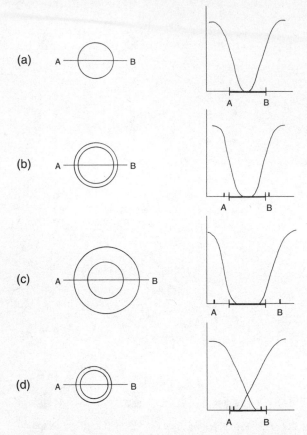

Fig. 4.6 The edge effect, and importance of buffering reserves to counter intrusions by exotic species. Left column shows the reserves, the lower three with a buffer zone (outer circle). Right column shows possible effects of this in countering invasion by exotic species; abscissa denotes diameter of reserve (A, B), ordinate denotes incidence/abundance of invasive species across this. (a) Simple reserve, no buffer; external species may invade from periphery leaving only central region pristine. (b) Reserve with narrow buffer zone added around core, reducing the effect of external invasion and resulting in a larger effective reserve. (c) Larger buffer, effective in ensuring the integrity of the core reserve. (d) Buffer established as part of existing reserve, so that core area is no longer large enough and the whole reserve is susceptible to invasion.

is somewhat oversimplified because the buffer zone itself may acquire additional species which are not wholly characteristic of the reserve or of the surrounding disturbed region.

All too often, the trend is the reverse of the above—rather than augmenting the original reserve size to create a buffer zone, part of the reserve is itself transformed to an intermediate status, so that the effective reserve is reduced in size or, even, the reserve rendered ineffective (Fig. 4.6d), as no part of it is protected from invasion.

The principles shown in Fig. 4.5 (Diamond 1975) have general applicability in reserve design but the details clearly differ from case to case. In general:

(1) minimizing the perimeter length : reserve area ratio decreases the potential for intrusion;

(2) larger reserves are preferred over smaller ones; and

(3) arrangement of separate reserve areas to maximize chances of passage between them is advocated strongly to avoid population isolation.

Provision of 'habitat corridors' is highly desirable, as many invertebrates will not traverse even small distances of inhospitable habitat (p. 93).

A further aspect which has engendered much debate is that commonly known as 'SLOSS' (Single Large Or Several Small), referring to the relative merits of a single large reserve or a number of smaller ones which constitute an equivalent area. The former has been recommended over a long period, again on the basis of island biogeography theory. Simberloff (1986) emphasized that smaller sites generally have higher extinction rates than large ones, and that there is indeed a critical minimum area needed by each species in order to support a viable population. Decline beyond that limit inevitably results in extinction, but the absolute size of a reserve may be small for many invertebrates, as noted earlier. Usher (1989) queried whether any minimum size for a nature reserve could be defined, and concluded that the answer may depend entirely on the taxa concerned. The 'several small' areas approach might be better for many invertebrates, depending on the degree of heterogeneity in any of the available options in

(1) each harbouring an independent and different assemblage; and

(2) serving to replicate discrete populations of notable taxa, rather than to conserve just one of these in a large area.

Simplistically, this could help safeguard against catastrophes which could wipe out a single site, and also provide founding material for possible re-introduction there later (p. 108). Environmental heterogeneity is universally relevant: the effective reserve for an ecologically specialized invertebrate species is the habitat with its specific resource needs, not simply the area in which that habitat is contained.

Evaluation of areas for reserves, and ranking any series of sites, is a complex process, and many different criteria have been employed for this (Usher 1986). A prime aim, for invertebrates, is to ensure that sites of special 'biological significance' are protected. The presence of notable taxa, singly or as critical faunas, can *per se* confer 'significance', but other attributes include a range of site-based and species-based features. 'Representativeness' is important, because it is presumed that nature reserves *will* conserve the vast number of invertebrates that are cryptic, undocumented, and even unknown in a template of biological diversity available for future investigation. Ranking of potential reserve areas for priority thus commonly evokes species characteristics.

Disney (1986a) noted that probably only four of the common criteria used in and site evaluation or ranking (Table 4.4), namely naturalness, rarity, area, and diversity, are really applicable for use with invertebrates, because of lack of knowledge of most invertebrate faunas. Three of these features are difficult to apply universally. 'Naturalness' implies comparison with some standard which may not be definable clearly, and often becomes a subjective appraisal of the degree of change which has occurred already. 'Rarity' necessitates knowledge of distribution and abundance, ideally from comprehensive surveys; Disney (1986a) noted that, because most sites are likely to harbour a few rare species of invertebrates ('rare species are very common', McArdle 1990), ranking may depend on the *number* of rare species, rather than the simple *presence* of rare species. Objective definition of rarity is needed (p. 76), and

adequate distributional data for most species may not be available. Rarity often equates to 'notable' without definitive evidence to confirm the state. 'Area', discussed in part above, is important also in combination with diversity. Smaller sites often have more species than would be expected; as the area increases, the *rate* at which additional species are encountered starts to decline, but the principle of adding 'new' fauna is important in influencing priorities.

For butterfly diversity in British woodland patches, the numbers of species present were correlated positively with woodland area, and the (calculated) area needed to support all 26 species was 458 hectares (Shreeve and Mason 1980). However, one wood of only 85 hectares contained 22 of those species, and addition of areas of representation for the other species gave a minimum real area of only 180 hectares for the presence of all the butterflies.

Thus, series of smaller sites selected for complementarity might collectively support more species, and on a substantially smaller total area, than a single large site.

The criterion of 'diversity' is applicable universally, on several different geographical and ecological scales, and information derived from comprehensive surveys or standard sampling programmes provides for objective comparisons. However, interpretation may be difficult from numbers alone because of lack of ability to distinguish (for example) resident and vagrant species, or native and exotic species.

Estimating diversity thus poses problems. It is difficult to collect, or sample, the total invertebrate assemblage even of a single site. For comparative ranking, the problems can increase very substantially, because it is necessary to sample at the same time each year, even the same time each day, in order to counter the effects of short apparency and of activity, or amenability to capture or detection. Comparative estimates of diversity necessitate comparable sampling programmes, including consideration of sampling effort and intensity, and details need to be tailored to the habitat and the kinds of organisms involved. A marine rock-shelf community, with a high proportion of long-lived sedentary invertebrate taxa is very different to evaluate from a grassland or forest community with a high proportion of mobile, winged insects, many of which may be conspicuous for only a few days or weeks each year.

Many different criteria have been used for comparative ranking of sites for reserve status in terms of the species present (Adamus and Clough 1978, for example) and many protected areas now incorporate consideration of two main parameters:

(1) values of the site itself, including topography, historical, and traditional values; and

(2) values (loosely, biological significance) accorded by the biota present.

These may direct priority for a particular site. Site or habitat evaluation thus incorporates values from the taxa present, but invertebrates have usually played small part in this, other than sporadic reference to a few well-known groups such as butterflies and dragonflies (Usher 1986). Factors such as 'endemism' or 'being type localities' are sometimes incorporated, but more emphasis has traditionally been placed on diversity.

Conservation value of sites can be assessed or ranked by attributing scaled values to invertebrates present and summing these to give a score for each site. This can be done at various levels of grouping rather than, necessarily, species alone.

Several quantitative or semiquantitative methods developed for ecological monitoring parallel closely the needs in ranking natural communities, and could be employed in this related context with little change.

Thus, one widely used index for assessing freshwater communities was developed in Britain by the National Water Council (1981) for studies on water pollution and is known as the biological monitoring working party score (BMWP). This was developed to help overcome the logistic difficulty of total species inventory and necessitates:

(1) determining and listing the families of macroinvertebrates present at a site;

(2) allocating numerical scores to the families; and

(3) summing these to give a cumulative site score.

The scores allocated to each family reflect the distribution, abundance, and ecological sensitivities of each, assessed on a wide (national) basis. Thus, several mayflies, stoneflies, and caddis flies likely to disappear under relatively low levels of pollution are ranked highly (each

Table 4.4 Criteria used in ranking and evaluating sites on wildlife values and others for conservation priority (partly after Usher 1986)

Diversity	Applicable to habitats and/or species
Naturalness	Implies sound 'baseline knowledge'
Rarity	Needs to be based on objective measurement
Area	Reflects susceptibility to disturbance
Threat of interference	May imply urgent need to prevent human interference
Representativeness ⎫ Typicalness ⎬	Is the site a 'good example' of the ecosystems it includes?
Scientific value	Broad range of interest
Educational value	
Ecological fragility	
Scarcity/uniqueness	Include remnant habitats
Archaeological or cultural interest	May imply that already modified from pristine condition
Importance for migratory vertebrates or others	These may act as 'umbrellas' for invertebrates

'10'), whereas the Chironomidae (p. 28) are ranked '2', and the class of segmented worms, Oligochaeta (often difficult to separate into families by the non-specialist) are given '1'. A high score for a site therefore represents a high diversity of families, and a high proportion of sensitive ones.

It may not be necessary to enumerate all families of larger invertebrates to rank habitats adequately. Green (1989) found that incidence of caddis flies (Trichoptera) alone gave a good indication of the conservation value of a series of English gravel pits; the aim was to determine whether the conservation values could be so assigned to unmanaged pits by comparing them with managed pits. Managed sites generally scored much higher than unmanaged sites, and the series of unmanaged pits could be ranked in order of conservation value by appraising the ecology of the caddis flies there.

Species level rankings are obviously more costly to obtain, but similar principles can be applied—of scoring each species, either as present or absent, or as a level of abundance from 'present' or 'few' to 'very abundant'.

As one example, again for freshwater assemblages, Chandler (1970) used five-minute sampling periods as his community content, and scored the abundance of each species, giving different values to pollution-tolerant and pollution-intolerant taxa. Each abundant, tolerant species would gain a very low score, but each abundant, sensitive species, a high one. Thus, presence (1 or 2 individuals/sample) of a highly sensitive species could gain a score of '90', whereas a very abundant (>100 individuals/sample) or tolerant species may receive a score or only 2–8, depending on the particular taxon. The ranking sequence for the Chandler index is subjective and amenable to change for particular needs.

In general, reserve selection may have two sequential components. The first involves descriptive ecology—identifying optimal areas on species richness, representativeness, or other criteria—and the second indicates how the long-term integrity of the reserve may be affected dynamically by internal and external processes (Webb 1989).

Reference sites

Because of the importance of deriving inventories or complete species lists of invertebrates for any site during a short period, and with limited resources, and the progressive demand or need for such information in evaluation and ecological interpretation, Disney (1986a) advocated the use of 'reference sites'. These, probably in many cases sites already reserved or protected securely (usually on the basis of criteria not involving invertebrates), were to be used for attempts at comprehensive surveys involving long-term sampling programmes. Such surveys are important in helping to assess the validity of shorter samples used for evaluation elsewhere, and also in helping to answer some more fundamental questions about the magnitude of biodiversity. May (1988), for example, commented that questions about this could be answered only by selecting a number of sites (including tropical forest sites), perhaps only a hectare in extent, and directly counting what was there. A similar approach was advocated by Di Castri et al. (1992) (p. 156).

The various problems in validating short-term surveys for invertebrates can, in part, be overcome by incorporation of standardized comparison from a reference site. Disney (1982, 1986a) proposed that a reference site should be included in any suite of sites it was desired to rank in some way. In each particular survey the mean numbers of individuals in samples from each site are divided by the mean for the reference site. An example is shown in Table 4.5, which illustrates the comparison of three woods with a reference site. This procedure assumes that the mean number of species/sampling unit is a valid measure of diversity for ranking consistently.

Invertebrate diversity in an isolated habitat may depend on the structure of surrounding areas, as well as the 'internal' parameters and the degree of isolation from other, similar, habitat patches. Webb (1989) examined the invertebrates of 22 patches of heathland in southern England. These patches ranged in size from 0.1 to 476 ha and were comparable in major features but had different degrees of isolation. Following

Table 4.5 Evaluation of faunal diversity by use of a reference site, using Diptera in British woods, from Disney (1986*a*)

Survey date	Wood	Mean N (Diptera species)	Rank value
June	X	29.8	186
June	Y	24.1	151
June	C	16.0	100
July	Z	43.2	129
July	C	33.3	100

Four woods were compared using data from two surveys on different occasions (June, July). C was the 'reference site', included in both surveys for comparison with woods X, Y, Z. Note that the mean number of species/sampling unit doubled at C in successive months, but rank values still enable comparison of Z (July) with X, Y (June). Mean of C catch was divided into means of the other sites on each occasion and results multiplied by 100 to give rank values expressed as a percentage of the reference site, and rounded to the nearest integer. *P* values were calculated from Mann–Whitney U-tests for each data set (June, July); in June, X was more diverse than Y, and Y more diverse than C; in July Z and C also differed significantly ($P <0.02$). In July, despite Z having a higher mean value of species number, the rank value implies that the fauna might have been less than that of X or Y.

inferences from earlier work on the same system (Hopkins and Webb 1984; Webb and Hopkins 1984; Webb *et al.* 1984), the area of a particular heathland and the area of heathland within a 2 km radius (taken as a measure of habitat isolation) were poor predictors of the diversity of beetles, heteropteran bugs, and spiders (Table 4.6). The various other habitats surrounding heaths (Fig. 4.4, p. 83) were correlated with heathland and, where more structurally diverse vegetation surrounds were present, there was some tendency for there to be greater invertebrate richness on the heath. Greatest increase

in richness occurred where deciduous or coniferous woodland and aquatic habitats were present nearby, and only for grassland was there *no* increase in heathland invertebrate richness.

For spiders, in particular, there was a downward trend in richness from woodland, through scrubland, to grassland and bare ground, thus with the diminishing structural diversity of the habitats.

Processes affecting a habitat patch or site therefore depend on characteristics of surrounding regions, as well as the area involved. An important general point, emphasized by Webb

Table 4.6 Invertebrate fauna of heathlands in Dorset, UK: correlations between species richness and site area (ha) and the area of heathland within the surrounding 2 km of a site (a measure of the degree of isolation) (from Webb 1989; method described in Webb and Hopkins 1984)

Taxon	No. species	Site area	Area of heathland within 2 km
Total Coleoptera	272	−0.41*	−0.58**
Phytophagous Coleoptera	65	−0.59**	−0.70***
Total Araneae	158	−0.07	−0.26
Heathland Araneae	60	0.38*	0.24
Heathland Heteroptera	15	−0.27	−0.46

Significance levels: *$0.05 > P > 0.01$; **$P < 0.01$; ***$P < 0.001$.

(1989), is that habitat patches (or, reserves) are part of a matrix over which an animal's probability of survival differs from location to location, with an *interchange* between the habitat patches and their surrounds. 'Patch dynamics' are rarely known in sufficient detail in any particular case, but emphasize the ecological complexities involved in designation of reserves on any arbitrary basis. This, and other such incidences, impinge directly on reserve selection and the subsequent need for management in order to conserve the characteristic invertebrates. Peripheral habitats in highly disturbed areas can play important roles as 'reservoirs', and the principle of 'more complex surrounding vegetation' is employed directly in some aspects of invertebrate habitat management (p. 83), for example in hedgerows and agricultural fields.

MANAGEMENT AND ECOLOGICAL UNDERSTANDING

In general, conservation management has three major contexts, which are not mutually exclusive:

(1) protection of species, communities, or sites against threats, including anthropogenic and natural changes;

(2) restoration after the impact of those threats; and

(3) control or regulation of threatening processes and factors.

Any, or all, of these are likely to be difficult to undertake to ideal levels, and in practice management is often a compromise arrangement, sometimes between controversies engendered by energetic dispute from different interest groups and protagonists. Management to conserve particular species will necessitate steps, based on detailed autecological study of such species, to assure the continued presence and accessibility of essential resources, such as particular foodplants, mutualistic organisms, and microclimates. Examples are given in Chapter 7. There is often a need for some form of anthropogenic change to enhance or supply such resources in reserves, and otherwise to increase

biological or ecological significance in various ways. Management may involve such processes as introducing stock of invertebrates (Chapter 5) in some instances, and, in addition to ecological processes, includes important regulatory components. These will be considered more in Chapter 6, and this section exemplifies some ecological modes of management and restoration for invertebrate benefit.

Managing modified landscapes

Management, then, includes deliberate creation of new habitats, through control of seral succession, and sometimes maintaining interference and external processes as a tool in achieving this. It can be directed at the habitat or site, usually to maintain it in, or restore it to, optimal condition to support high diversity or 'typical' representative communities or assemblages; or at species, to improve or augment a particular suite of resources for a given taxon. In many instances, management can be implemented only in reserves, but dramatic changes in protected areas may be viewed with suspicion by concerned people and good 'public relations' may be needed to avoid conflict of ideals. However, many active management steps for invertebrates are necessarily in habitats in disturbed landscapes, such as agricultural land, forestry plantations, and seashores. As well as increasing or conserving biological significance, much management also has an amenity component, or more tangible value, such as conserving natural enemies of crop pests, or arthropod food for game birds, on field margins (Dover 1991). More embracingly, aspects of landscape ecology, with consideration for integrating the elements of 'corridors', 'matrices', and 'patches' in relation to invertebrates (Samways 1989*b*, and included references), deserve appraisal in examining the integrity of any site in relation to invasion by other species or imposing topographical barriers. These may, in essence, determine the distribution of taxa. Management knowledge thereby becomes an integral facet of reserve/habitat design, and the shape, size, and features of available habitat usually dictate or influence the scope for conservation management.

Table 4.7 Relative impacts of 'corridors' on nine species of bush crickets (Orthoptera: Tettigoniidae, Decticinae) near Montpellier, France (Samways 1989a)

Species	Kind of corridor				
	Untarred road	Tarred road	Roadside verge	Drainage ditch	Stream or river
Decticus albifrons	–	–	6	6	3
D. verrucivorus monspeliensis[a]	2	2	8	5	1
Platycleis affinis	–	–	7	5	3
P. albopunctata	3	3	–	5	2
P. fedtschenkoi azami[a]	2	1	5	8	1
P. falx	–	–	6	6	3
P. intermedia	–	–	7	5	3
P. sabulosa	–	–	5	5	3
P. tessellata[b]	2	2	6	5	1

Barriers: –, no or very minor barrier; 1, almost complete barrier; 2, partial barrier; 3, minor barrier.
Habitats: 5, marginal; 6, refuge; 7, preferred; 8, only suitable.
[a] Wingless species.
[b] Flies only short distances, < 5 m.
Other six species are stronger fliers.

Corridors and habitat continuity

Diversity of the effects of corridors, and the difficulty of extrapolating to generalize about these, is illustrated well by Samways' (1989a) comparative study of nine species of bush crickets (Orthoptera) in southern France (Table 4.7). The extent of a barrier's effect differed considerably for different taxa, depending on the mobility of the particular species. The impact of different corridors on different species ranged from highly important (sometimes as their sole habitats) to very hostile environments which could retard interchange of individuals from each side.

The width of a corridor can influence strongly the amount of dispersal that an invertebrate will undertake voluntarily. Baur and Baur (1990, 1992) monitored movements of a terrestrial snail, *Arianta arbustorum*, in Sweden. In a 3 month period, only 10 (of 84) snails crossed an overgrown path only 0.3 m wide from the adjacent verge. Two (of 194) snails crossed a 3 m wide gravel track used by about 200 people/day, and only one crossed an 8 m wide tarmac road over the same period. The latter barrier may be founded in biological reality: insufficient mucus

production could lead to desiccation of the snails during the trip, and the tarmac is also likely to be devoid of stimulants such as food odours. This example illustrates well the likelihood of inadvertent habitat isolation from apparently trivial intrusions such as a walking track.

Roads are 'artificial habitats', unfamiliar to many species (Mader 1984). As with the snail noted above, some ground beetles (Carabidae) will not cross roads (Mader 1984). If placed beyond the verge, they return to it rather than cross the open space confronting them, as did all examples ($n = 30$) of the lycosid spider *Pardosa amentata* tested in this way (Mader *et al*. 1990). However, even relatively natural areas can prevent dispersal.

Some butterflies are reluctant to traverse even a few metres of open ground if this is an inhospitable habitat (Ford 1975). 'Barriers' of this nature are a real biological phenomenon and their effect may need to be studied in planning management strategies for any given invertebrate species. Management may be needed to overcome their population-fragmenting function. Mader *et al*. (1990) noted that the extra movement sometimes imposed

Table 4.8 Suggestions for management to enhance butterfly richness on new road verges in Britain (after Munguira and Thomas 1992)

	Suggestion	Rationale
1.	Make verges as wide as possible	Wide verges contain a greater variety of breeding habitats; support higher densities of butterflies
2.	Improve, by making additional habitats	Change to support a wider range of species: this is often well below potential
3.	Permit vegetation change, e.g. by reducing fertilizer and topsoil applications selectively	Produce sward with greater potential for other plants to invade; may be necessary to sow specific larval foodplants for selected taxa
4.	Promote irregular topography, such as by step-cutting slopes, for different or adding ditches	Produce a greater variety of breeding habitats, with a range of growing conditions foodplants
5.	Where possible, plant bordering hedgerows, or clumps of shrubs and trees	Foster diversity; break up uniform stretches; shelter; increase breeding habitats
6.	Rationalize mowing regimes	Rather than mowing mid-summer (usual to increase visibility), mow only a relatively narrow strip then, with more extensive winter mowing. Ideally, maintain diversity through a mosaic of mown/unmown patches. Mowing is a major disturbance and causes a substantial decline in butterflies, with slow recovery

by the presence of barriers may decrease the effective dispersal range of invertebrates and lead to exhaustion of energy reserves vital for use on reaching a new habitat patch. Propensity for dispersal differs substantially between different groups of invertebrates. It is a 'non-event' for permanently sedentary forms such as barnacles once they have passed the pelagic juvenile phase, but, conversely, is theoretically simple in certain winged insects that have the physical capability to fly long distances. For conservation of many invertebrates, it may be necessary to construct corridors, or to maintain those developed primarily for other reasons. Roadside plantings of trees, for example, for windbreaks, landscaping, light shields, erosion control, and numerous other reasons, may also be important as refuge habitats for animals and plants. Indeed, populations of some insects have been reported to become very abundant, as outbreaks, on roadside verges (Lepidoptera, Hemiptera) (Port and Thompson 1980; Port and Spencer 1987), possibly reflecting the high nitrogen content of roadside vegetation and its nutritional suitablity for such herbivores.

From their recent study of use of road verges in southern England by diurnal Lepidoptera, Munguira and Thomas (1992) found that roads themselves were not a barrier to gene flow of any species in their study. Up to 30 per cent of adults of the three butterfly species with closed populations (*Maniola jurtina, Melanargia galathea*: Nymphalidae: Satyrinae), and *Polyommatus icarus* (Lycaenidae)) crossed roads, and little mortality occurred to any butterfly species from vehicles: only up to around 7 per cent in some open-population species.

Munguira and Thomas (1992) recommended a series of management steps that could enhance the value of road verges as butterfly habitats; these are noted here (Table 4.8), not only as a specific and well-documented case, but also to illustrate the range of complementary factors that might be involved in any invertebrate habitat management programme. The authors point out that road-verge area in the UK is around 212 200 hectares, roughly equivalent to the total land area of the country's nature reserves, so that it is not in any way a trivial habitat. The principles involved in managing road verges for invertebrates in the UK are:

(1)　to reduce habitat uniformity;

(2)　to increase 'naturalness'; and

(3)　increase the diversity of breeding habitats, especially through fostering plant diversity for this specific group of herbivores.

Each of the steps in Table 4.8 is likely to benefit other invertebrates as well.

Assemblages

Problems can arise from management when it is targeted for a number of different notable taxa, with different needs, occupying the same site. A high proportion of the small British butterfly fauna frequents chalk grassland, and a number of species are confined to this habitat. Much chalk grassland has been destroyed, so there is now a mosaic of sites, many of them recognized as floristically or otherwise important, which are managed, or their use influenced by conservation bodies. One consequence of reservation has been that traditional (historical) management practices, such as grazing by domestic stock, has sometimes been curtailed. However, grazing is indeed necessary to maintain the habitats needed by some butterflies, and regulated grazing regimes are now accepted as an important component of grassland management for these and other invertebrates (BUTT 1986). Likewise, control of rabbit populations by myxomatosis in the 1950s markedly reduced their grazing impact. These measures have shown that, for some ecologically sensitive butterflies (particularly Lycaenidae), there is a very fine balance between 'overmanagement' (for example, by overgrazing when some grazing is indeed needed to maintain grassland or herbs) and 'undermanagement' (such as total lack of grazing needs). The adonis blue, *Lysandra bellargus*, breeds almost solely in areas where the larval foodplant (a vetch, *Hippocrepis comosa*) grows in turf less than 3 cm tall, especially 0.5–1 cm; the brown argus, *Aricia agestis*, needs turf about 1.5–5 cm high; the common blue, *Polyommatus icarus*, is rarely common in short turf areas but prefers around 4–10 cm, as examples of differences, and the competing needs of these species (and others) dictate management for a regime to maintain a suite

of different conditions on the same site. The management options for this system, discussed by BUTT (1986), include:

1.　Grazing regime: continuous stocking, rotational stocking, seasonal grazing or sporadic/opportunistic grazing.

2.　Kind of grazing: by cattle, horses, sheep, rabbits, or others.

3.　Mowing, viewed as a poor alternative to grazing because results are often unpredictable, but used as a restricted option in areas where grazing is impracticable due to such factors as fencing costs, small area, inaccessibility, or mosaic needs.

4.　Burning, favoured occasionally to control build-up of litter, but believed to be highly damaging to some invertebrate groups (such as molluscs) (Kerney and Stubbs 1980). It can help to maintain floristic diversity.

5.　Scrub control: scrub can be of considerable importance in providing habitat edges, roosting sites, and foodplants; and management needs to maximize the benefits of shelter and edge effects by controlling the kinds of scrub present and their distribution, preventing excessive invasion of grassland areas.

6.　Many chalkland butterflies may be benefited by bare ground, because some larval foodplants colonize *only* bare ground. Very localized scraping off of turf to facilitate pioneer community establishment may be needed occasionally, and is clearly an option for creating habitats in this system.

Such programmes for butterflies have been discussed extensively (for example, Fry and Lonsdale 1991; New 1991a; Thomas 1991) and include some of the classic advances in understanding of conservation needs of invertebrates. They are complicated by the need to provide for two very disparate life stages: caterpillars and butterflies have very different ecological requirements. Habitat management for temperate-region butterflies has advanced to a more sophisticated level than for most other invertebrates, but many cases manifest one of two general situations.

(1) the need to control natural successional processes; or

(2) modifying treatment of a particular vegetation to maintain it in optimal condition, not necessarily as a component of succession.

Broader appraisal of chalk grassland invertebrates (Morris 1971, 1979), and a wide series of approaches to managing a range of different British vegetation systems and other habitats, such as wetlands and coasts, for invertebrates (Fry and Lonsdale 1991; Kirby 1992) are exemplary as a basis for assessing needs for planning in other parts of the world.

Although much of the ecological understanding that underpins effective management of invertebrates has come only in recent years, the principle of managing some natural invertebrate resources is not new, but has usually been undertaken more through regulation (p. 114) than primary 'ecology'. However, much 'traditional' resource management has indeed fostered well-being of invertebrates, especially in terrestrial systems.

Invertebrates in agricultural systems
Both maintenance of succession and the re-creation of habitats by purposeful restoration may need to incorporate the kinds of consideration exemplified above. Consider, for example, the needs of invertebrate conservation in intensively managed industrial areas in Europe, that now typically support many species that depend on sympathetic farming practices for their long-term survival in highly modified open landscapes (Fry 1991). The detailed studies on invertebrates in the British agricultural landscape illustrate well the ecological basis for effective management in modified regions.

Agricultural expansion following the Second World War led to the removal of many invertebrate habitats, such as hedgerows, in the interests of production and harvesting efficiency, with little appreciation for the roles these played in the ecology of the agricultural landscape, as refuges for invertebrate diversity, and as

habitats that maintained reservoirs of the natural enemies of many important crop pests (Lewis 1969; Pollard *et al.* 1974). This trend to consolidation of smaller land units resulted in removal of a high proportion of hedgerows and ditches, with important impacts on the associated fauna. Since about 1960 (Fry 1991), three main changes have occurred in the agricultural landscape:

(1) lowered heterogeneity;

(2) lowered diversity of the component elements; and

(3) lowered contact between neighbouring habitats.

In addition to the value of corridors (above), the intersections of these ('nodes': Forman and Godron 1986) are also of great importance. Nodes, the junctions of field boundaries and the like, function as:

(1) intersects that increase the connectivity between landscape points and habitats; and

(2) distinctive habitats *per se*.

Fry (1991) showed that the diversity of invertebrates at nodes in agricultural systems may be greater than that at the mid-points of boundaries, probably because of greater habitat diversity there and lower disturbance from machines and chemicals. Enhancement of the conservation value of field boundaries, and appreciation of the value of headlands (the area between the boundary and crop, which is used for turning machinery) are important developments in managing agricultural systems for invertebrate conservation. Field margins have been promoted in Britain for harbouring arthropod food vital to gamebird chicks. Dover (1991) reviewed relevant management steps, especially for 'conservation headlands'—the unsprayed outer 6 m of cereal crops, the distance reflecting the length of a normal tractor-mounted spray boom.

As well as benefits to butterflies, which have been studied extensively (Dover 1989, 1990; Dover *et al.* 1990, for examples), the overall habitat available for invertebrates is increased massively. For example, a headland each side of a hedgerow effectively widens a 2–3 m corridor by a further 12 m, with provision for many additional nectar plants, larval foodplants, and 'refuges' (Boatman *et al.* 1989).

Parameters for establishment of conservation headlands, which receive reduced and selective pesticide applications, include:

(1) no wide-spectrum herbicides against broad-leaved plants;

(2) no insecticides after 15 March in the year of harvest;

(3) selected grass herbicides; and

(4) exclusion of some fungicides that also have invertebrate activity.

In order to prevent weed invasion from the boundary into the headland (crop), a 1 m sterile strip is created between them by broad spectrum herbicide (Fig. 4.7). The headland serves also to reduce spray drift on to boundaries, and reduced fertilizer impact is also being investigated to prevent loss of plant diversity through eutrophication. Fry (1991) noted that some 730 km of conservation headland exists in Britain, with the maintenance costs paid by the farmer-owners. Further investigation is needed on the roles of these habitats for invertebrate conservation, as noted more generally for many aspects of field boundaries by Morris and Webb (1987): they emphasized that field margins 'will continue to support a wide range of species and to provide habitats, or habitat components, for them'. The overall importance of field margins in relation to reserves and other protected areas needs further evaluation.

Unsprayed headlands can support significantly higher densities of non-pest arthropods than herbicide-treated ones (Chiverton and Sotherton

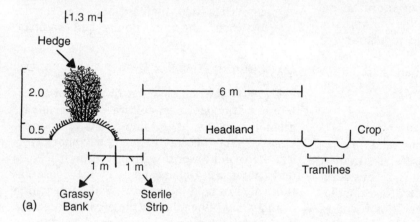

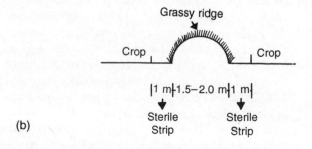

Fig 4.7 Hedgerow/ridge management in British agricultural systems: (a) the principle of conservation headlands; (b) grassy ridges established across fields to act as reservoir habitats and facilitate entry of arthropod predators to crop environments. (After Dover 1991; Thomas *et al.* 1991)

1991), together with more weed species, higher weed densities, greater weed biomass, and higher weed cover. There were also higher densities of predatory arthropod groups, especially polyphagous species (such as earwigs and ground beetles). These areas constitute important refugia for such animals, important as predators of crop pests. Recent studies (Thomas *et al.* 1991, 1992) have indicated how access of these predators to crops can be improved by providing 'island corridors' during normal cultivation practices. Careful two-dimensional ploughing in autumn can be used to create ridges about 0.4 m high and 1.5–2 m wide across fields, leaving about 25 m each end to allow for machinery passage to maintain the field as a single cultivable unit. These ridges are sown with grass (recommended: *Dactylis glomerata* at 3 g/m^2 or *Holcus lanatus* at 4 g/m^2, or a 50 : 50 mixture of these two), and these areas are meant to mimic (and replicate) the grassy area typically left flanking a hedgerow (Fig. 4.7). The ridges become effective after 2 or 3 years and provide overwintering habitats for spiders and predatory insects. They provide them with easy access to the crops and increase levels of crop protection—especially against aphids on wheat. This method is being promoted to farmers as a cost-effective control method without interference to normal farming practices, in conjunction with reduced pesticide use.

In terms of conservation management, the technique has demonstrated the capability of 'beneficially manipulating the available environment in an ecologically short time' (Thomas *et al.* 1991). Such cultural techniques are likely to become more widespread.

Monitoring 'ridges' over 3 years showed progressive change in the composition of spider and ground beetle faunas, from pioneer to more specialized species, as the habitats matured (Thomas *et al.* 1992). The main change in spiders was in the balance between Linyphiidae and Lycosidae, with the latter being more permanent and specialized species (Huhta 1979, Nentwig 1988). The number of boundary carabids in relation to open field taxa also increased over this period.

Succession

Usher and Jefferson (1991) commented that 'ecological succession is perhaps the single most important factor causing change in the arthropod community'. The examples noted above indicate well the need to manage many habitats to maintain particular successional stages. Any management for invertebrate conservation may be combined with that for other priorities: forest management for gamebirds in Britain benefits insects (Robertson *et al.* 1988); cutting and burning are used to manage reed beds for reed harvesting (Ditlhogo *et al.* 1992); heathlands in southern Australia may be control-burned to restrain succession to *Banksia* woodland, and to reduce fuel loads to curtail the impact of sporadic accidental bushfires, for example. The reed example involves invertebrates that can damage reeds, and which are economically important when reeds are used for thatching, and some taxa of conservation significance. The Australian heathland case exemplifies those where characteristic communities (including rare taxa) may be displaced, and also influences adjacent seasonal wetlands that support other notable taxa: the damselfly *Hemiphlebia mirabilis* at Wilson's Promontory National Park, Victoria, is discussed in Chapter 7 (p. 135). Use of fire in management is controversial, and several geographically and ecologically discrete studies showed that burning can reduce populations of soil invertebrates. It is, however, difficult to generalize. High-intensity burning of heath and eucalypt woodland in semiarid Victoria led to *doubling* of the number of ant species on the ground and to marked changes in community composition, by reducing previously dominant species (Andersen and Yen 1985).

Ditlhogo *et al.* (1992) suggested that few soil or litter invertebrates were, in fact, affected by normal reed management, and that the effects on above-ground species were transient. Some representatives of both groups increased in abundance, and heat from burning in the saturated reed-bed soils had no major effect on the invertebrates. Burning of dry reed beds might have more severe effects and, as in many other instances of successional management, mosaic treatment (here, of rotational burning or cutting and leaving undisturbed) is likely to minimize long-term adverse effects by facilitating chances for recolonization of treated areas, should this be necessary. A mosaic of many habitats is often needed to conserve a characteristic fauna of a particular vegetation type or water body. Diversity of South African fynbos ants (Donnelly and Giliomee 1985), for example, declined substantially after habitat burning, and representation of the 36 species involved requires a mosaic where different successional stages are represented.

Most work on successional influences on invertebrates has emphasized terrestrial arthropods (Usher and Jefferson 1991), but has demonstrated adequately that, without such management, many species characteristic of earlier seral stages are likely to become extinct. Details of the management needed will depend on defining particular 'conservation goals' and designing steps to achieve these. Usher and Jefferson noted that two distinct processes are involved:

(1) the need to retain a particular successional stage through preventing it from maturing to a later sere; or

(2) the need to re-commence a succession locally by designing and implementing regimes of 'disturbance'.

There is also a clear difference in emphasis, mirrored in the 'goals', between managing for particular rare (often, 'threatened': p. 169) taxa, and for diversity—the multitude of apparently unthreatened invertebrates that might benefit from any kind of habitat security or enrichment.

New habitats

The creation of new habitats for invertebrates takes many forms. Three examples are given, to illustrate the diversity of needs, and to draw on experiences amenable to replication and parallel in various parts of the world. However, any such operations should be planned very carefully. Kirby (1992) warned that established conservation interest should never be put at risk by 'conserving' what is not yet there. In the past, many invertebrates have suffered from misguided management, simply because their habitat needs have not been recognized.

Ponds for dragonflies

The British Dragonfly Society (1992) has advocated strongly the conservation advantages of 'digging a pond for dragonflies' and also noted the desirability of renovating old ponds that may have been rendered unsuitable by having dried out, been used as rubbish dumps, or been neglected and allowed to become overly shaded, for example. If rehabilitation is followed, monitoring of the dragonfly population before and after this is recommended to determine the efficiency of the changes made. Restorative steps needed might include:

(1) clearing, in stages over several years, of excessive vegetation, without use of herbicides;

(2) removal of excess silt, again in stages, and attempting not to disturb margins unduly;

(3) increasing exposure by removing or pruning nearby shrubs and trees;

(4) controlling nutrient input, such as manure in farm ponds, and countering pollution from insecticides or fertilizers from washing out of farm machinery or spraying equipment.

New ponds should be as large as possible; in Britain they have a minimum viable size of around 3.75 m², the small size indicating, yet again, that many invertebrates do not necessarily require large areas in which to live. Design considerations should include:

(1) shelter from prevailing winds;

(2) not to be shaded excessively by overhanging trees;

(3) avoiding run-off from roads or intensively farmed areas;

(4) variations in depth, with the deepest part a minimum of 60 cm (small ponds) or 75 cm (larger ponds); and

(5) avoiding vertical sides and sharp corners if possible, and including some shallower

margins to foster the growth of emergent plants.

Physical changes to create desirable microhabitats

The adonis blue butterfly in Britain is on the northern fringe of its European range, and can thrive only on south-facing isolated slopes in the south of England (Thomas 1983), where its caterpillars feed in very short swards. Thomas suggested that additional suitable habitats could be created by use of earth-moving equipment and explosives, and that such augmentation could be important for the butterfly's survival. Use of such management tools in nature reserves might be difficult to accept, because of other interests involved, and even the suggestion might lead to controversy.

Artificial reefs for marine invertebrates

Artificial reefs have been constructed in many places, mainly for amenity values in increasing local populations of fish for recreation and tourism. Materials used include rocks or concrete waste, either 'dumped' in piles or arranged more carefully in 'underwater cairns', and old vehicle tyres piled and tied together. These, and others, constitute substrates for colonization by animals and plants, with holes serving as refuges for larger animals, including cephalopods. Their use as an invertebrate conservation tool has not been explored deliberately on any wide scale, but seeding of such reefs (for example, by coral transplantation) (Harriott and Fisk 1988) may sometimes be practicable.

Replacement habitats for many invertebrates, in some sense new, are provided by exotic plants such as forestry monocultures, or from increases of any other exotic resource which is amenable to exploitation by native consumers. Other than feeding directly on such plants, management may need to incorporate consideration of corridors, edges, shade regimes, soil pH changes, and other (mainly physical) aspects of the environment. Older woodland management in Britain, coppicing, tended to foster a high diversity of butterfly species, for example, and decline of such practices, with consequent increase in shade, led to their decline in many

woods (Heath *et al.* 1984). Woodland clearings and rides are important butterfly habitats (Warren 1985), and increasing shade in conifer plantations as the trees mature is likely to induce many local extinctions, without an accompanying programme of ride-widening and establishment of glades and open spaces (Greatorex-Davies *et al.* 1993). For instance, rides of 30 and 40 m width are required as the conifers reach 20 and 25 m height, respectively, in order to maintain light levels sufficient for the rarer butterflies. Without provision for this, shade-intolerant butterflies and other taxa are likely to decline.

Habitat restoration may have tangible economic effects for people who depend on invertebrates for their incomes, and is often motivated by this concern. Blue crabs (*Callinectis sapidus*) form the basis of an important fishery in Chesapeake Bay, Virginia, and the crab's early stages use seagrass (*Zostera marina*) beds as a preferred habitat. Since the 1960s much *Zostera* has been eliminated by pollution, so that the density of crabs has decreased, and catch/effort ratio reduced (Anderson 1989). A combination of pollution control and seagrass planting to restore the habitat would lead to an estimated net economic benefit to the crab-fishing industry of around US$2.4 million. This figure reflects the increased supply of crabs and the consequent lower price. The economic benefit to the consumer is the difference between the maximum amount that consumers of crabs would be willing to pay for the quantity consumed, and the amount actually paid. Such figures clearly appeal to both producers and consumers, and Anderson's (1989) approach is valuable in demonstrating tangible benefits in this broad way. Long-term benefits accrue, but others (such as for commercial fish stocks) might also occur.

In summary, management or restoration of habitat is an integral part of conservation and is important for many invertebrates. There is often a range of acceptable options for this, with different consequences for invertebrates, and the 'best option' may be difficult to predict without prior knowledge. For example, opencast mining initially leaves large areas devoid of vegetation, with bare soil. The variety of future uses anticipated for such areas can result in minimal subsequent treatment (allowing natural succession to occur), covering with topsoil, seeding with grasses, or planting of shrubs and trees leading to more intensive afforestation or reafforestation. Comparison of development of spider and ground beetle faunas over 15 years on initially bare ground left alone or neighbouring afforested areas showed differences in species composition, but no significant differences in numbers of species present (Mader 1986). Forest species either were not found in the natural area, or occurred there in relatively low numbers.

Finally, an important aspect in determining the need for management is some provision for monitoring any notable taxa for which a reserve is designated specifically or as a priority, within the reserve. The endemic Tasmanian freshwater crayfish *Astacopsis gouldi* is, perhaps, the largest freshwater crayfish in the world (it reaches weights of 4–5 kg), but had apparently declined through fishing pressure to the state where it was considered 'vulnerable' by Wells *et al.* (1983; p. 120). Most specimens caught recently are much smaller than the maximum size, up to about 2 kg (Horwitz and Hamr 1988; Horwitz 1990a). However, consolation for the species' well-being had been drawn from the erection of a reserve specifically for *A. gouldi*, Carbine Creek Freshwater Crayfish Reserve (founded in 1968), in which the population was to be protected. Surveys in 1987 (Horwitz and Hamr 1988) revealed only low population levels at Carbine Creek, and *no* large mature crayfish were found. Large numbers of bait lines were present, suggesting that illegal fishing might have contributed to further decline, and that management of the area had been inadequate to achieve its primary purpose. The population at Carbine Creek was not considered to warrant preserved status, and Horwitz and Hamr (1988) urged that a new reserve (or reserves) be established with suitable management and adequate policing for this notable species. In addition, the legal trap limit for *A. gouldi* (12) was considered excessive. Horwitz (1990a, 1990b) recommended that this be reduced to three at the most, and that females should perhaps not be taken at all because it is suspected that they breed only once every 2 or 3 years.

More widespread protection for *Astacopsis* has potentially been provided by the declaration of a series of northern rivers in Tasmania as sanctuaries, and reduction of the catch limit elsewhere to two males a day. Although habitat degradation may still occur, these steps (operative from May 1993) suggest a more positive future for the world's largest freshwater crayfish.

FURTHER READING

Buckley, G. P. (ed.) (1989). *Biological habitat reconstruction*. Belhaven, London.

IUCN (1990). *The IUCN red list of threatened animals*. IUCN, Gland.

Kirby, P. (1992). *Habitat management for invertebrates: a practical handbook*. Royal Society for the Protection of Birds, Sandy, Bedfordshire.

Southwood, T. R. E. (1978). *Ecological methods, with particular reference to the study of insect populations*. Chapman & Hall, London.

Spellerberg, I. F. (1993). *Monitoring ecological change*. Cambridge University Press, Cambridge.

Wells, S. M., Pyle, R. M., and Collins, N. M. (1983). *The IUCN invertebrate red data book*. IUCN, Gland.

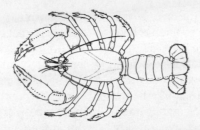

5 CAPTIVE BREEDING AND INTRODUCTION OF INVERTEBRATES

INTRODUCTION

The topics dealt with in the previous chapter relate primarily to detecting and implementing needs for conservation in the field. However, there is another aspect of increasing importance for invertebrates: so-called '*ex situ* conservation'. This incorporates the maintenance of captive-bred stocks and their release into the wild, either to augment dwindling natural populations or as some form of introduction or re-introduction to natural habitats. The process has stimulated much thought over optimal methodology for captive rearing, the situations in which *ex situ* conservation is needed or can be justified, and the need for understanding the genetics of the taxa concerned, and of the principles involved in maintenance of natural variability and its consequences. These topics are discussed in this chapter.

Captive rearing, or some related form of controlled breeding, of some invertebrates is undertaken for reasons other than conservation interest—mariculture of edible molluscs and crustaceans, for example, has long been an important facet of fisheries industries and 'butterfly ranching' to satisfy commercial demands for perfect specimens by collectors is also important. These, and others, although not associated directly with practical conservation, are important in the development of invertebrate conservation, because they have been instrumental in leading to the development of technologies,

rearing protocols, and accumulation of much detailed biological knowledge, which can form the basis for rearing related, non-commercial species which might be vulnerable or threatened. In other instances, it is likely that 'dual purpose' operations—such as combining commercial and conservation motives for insects—can be developed, for example by releasing a proportion of individuals of the rare butterflies reared for commercial sale. Likewise, much knowledge of rearing invertebrates has come from the need to maintain laboratory stocks of selected taxa for experimentation, teaching, or as food supplies for vertebrates, and technologies for rearing many of them on natural or semiartificial diets are well established. Mass-rearing of predators and parasitoids, as well as of herbivores, has been pioneered through their need for biological control releases. In short, although each species has particular requirements for successful captive maintenance, much 'generic methodology' for various invertebrate groups is available from non-conservation contexts as a basis for elucidating these.

The role of captive breeding of invertebrates for potential release to the wild, or for translocation to additional sites, is gaining recognition as a valuable (even, integral) facet of conservation programmes which can complement *in situ* conservation constructively. As field populations decline, they may need to be augmented regularly from laboratory-reared stocks in order to maintain their viability. The introduced subspecies of the large copper butterfly (*Lycaena dispar batava*) in the British fens, for example, has been augmented nearly every year since the 1930s, and latterly re-established in Woodwalton Fen from captive stock after it became extinct (Duffey 1977, 1993). This is one of the longest-term programmes involving this kind of management regularly and, without such expensive subsidy, the butterfly could not survive in the wild.

In other instances, laboratory stocks represent the sole surviving populations of taxa which are extinct (or believed to be extinct) in the wild. A number of species of *Partula* snails from Pacific islands, such as Moorea (p. 61), exemplify this. What were believed to be the last nine specimens of *P. hebe* were collected from the island of Raiatea in 1991 by an expedition from the London Zoo, and the captive population has increased in size substantially (Schoon 1992). It is hoped that this, and other species of *Partula*, will eventually be re-introduced to the wild once 'safe' sites, from which the predatory snail *Euglandina* (p. 60) has been eliminated, become available. There is little doubt that, except for this rescue effort, some of the species would already be extinct, and the future of some is, in any case, not assured. The philosophy behind such attempts has long been important in vertebrate conservation programmes, but is still relatively unusual for invertebrates. Part of this difference simply reflects that large mammals and birds have long been major attractions in zoos, and many people accept unhesitatingly the need for such measures as maintaining primates or rhinos in captivity to guard against their illegal exploitation in the wild. For most vertebrates, the priority for considering *ex situ* conservation post-dates accumulation of collections in menageries. For most invertebrates this traditional precursor does not exist, except in a few zoos with a long tradition of showing selected taxa.

CAPTIVE BREEDING

Several contributors to Collins (1990) emphasized that captive breeding of invertebrates has only rarely been of importance in conservation,

because most invertebrates are not threatened by overcollecting but, rather, by habitat loss. Without habitat restoration and continued management it might not be practicable to release the species into the wild at some later stage. However, as the likelihood of releases becomes more frequent, satisfactory protocols for rearing threatened invertebrates and maintaining them sufficiently fit for subsequent liberation need to be developed. These must incorporate aspects of husbandry and genetic management. The two cases noted here exemplify the progress that is occurring.

Partula *snails*

Decline and extinction of *Partula* in the wild has led to the situation where the only living representatives of many species are those held in captivity, as the 'Partula Programme', supported since 1986 by the Captive Breeding Specialist Group of the Species Survival Commission. The programme has been reviewed recently by Tonge and Bloxam (1991), and the following summary is from their account. In 1990, 15 zoological collections maintained *Partula* colonies, in Britain (seven collections), USA (four), Australia (two), Belgium and Germany (one each).

The intermediate levels of speciation in the genus, one of the reasons for its evolutionary significance and interest, have provided problems for captive breeding programmes in assuring the integrity of segregates selected for breeding. High mortality rates can occur, due to such factors as minor changes in diet when snails are transferred between collections, or in the physical environment (presence or absence of ultraviolet light, changes in daily temperature fluctuations). The poor breeding and survival rate in most collections have ensured that developments of husbandry methods have had to take precedence over development of a suitable genetic protocol (p. 106), merely to establish colonies.

A very high proportion of captive *Partula* (94 per cent of 4077 individuals) in 1990 were of two Moorean species, *P. taeniata* and *P. suturalis*, with another 11 species represented in the remaining 213 snails. Three species are listed by Tonge and Bloxam (1991) as now extinct in captivity, *P. exigua* (Moorea) and *P. filosa* (Tahiti), for example, have become extinct during the 1980s, and prospects for several other species were assessed as poor (with extinction likely in less than 5 years). The two species most successful in captivity were also the most widespread on Moorea in the past.

This programme is particularly significant because of the strong conservation motivation and need, where substantial specialist expertise was available, multiple sites for captive colony formation were used to 'spread the risk', and where results have been less than ideal. Despite widespread optimism that captive breeding would save most species of *Partula* from Moorea and Tahiti, this is clearly not the case. Without it, though, there is no doubt that they would be extinct already, and it seems likely that at least three of the Moorea species will persist for at least the medium term in captivity; there are plans to set up *Euglandina*-free enclosures on Moorea as release sites for re-introductions to be made.

The red-kneed bird-eating spider, Euathlus (Brachypelma) smithii

Concern for this large and attractive spider in the wild has arisen because large numbers are captured in Mexico to keep as pets (especially in the United States and Europe), and as souvenirs for tourists. Wells *et al.* (1983) recommended that trade should be monitored, and the spider is now included in CITES Appendix 2 (p. 118). Its natural range may also be declining due to disturbances to marginal agricultural land. This docile spider is probably not yet threatened seriously in the wild, but captive propagation is likely to be of considerable value in reducing the potential for overcollecting in Mexico. The exemplary breeding programme for *E. smithii* at the London Zoo (Clarke 1991) indicates the likely long-term viability of the spider, and the experience gained is likely to be transferable to other species.

Large numbers of young (up to around 700 juveniles/egg case) are produced, and growth is slow. It seems that immature stages take of the order of 5–7 years to reach adulthood, and females can live for 28 years.

A preliminary genetic protocol (p. 106) involves a foundation stock from 10 unrelated egg sacs, and 'coding' of spiders so that they can be recognized individually will ensure that each will be used only once for breeding. A maximum of 40 young from each egg sac will be retained for breeding, with the assumption that 20 will survive to adulthood and 10 of those will be used for breeding for the next generation. Turnover of generations in captivity will, of course, be slow—much slower than for most other invertebrates for which captive breeding may be contemplated. The spider, in common with many other predators, is logistically intensive to rear because spiderlings must be kept alone, each in an individual container, to avoid cannibalism. There is some possibility of being able to manipulate individual growth rate by changing the rearing temperature and food supply, and Clarke (1991) also noted that identification of the sex of spiderlings from characters of their cast skins may be useful in 'streamlining' rearing programmes by avoiding 'redundant rearing' of numerous individuals.

Several zoos are involved in the London-based rearing programme, together with members of the British Tarantula Society, illustrating the possibility of fruitful co-operation with amateur enthusiasts in such schemes. The 'problem' of likely overproduction of a species of conservation significance with attendant dilemmas of what to do with surplus stock, is highly unusual!

Both of these programmes include points applicable to many other cases, not least the problems of logistics (cost, efficiency) involved. However, for conservation of some species there may be no practical alternative to re-introduction from captive stocks, so that the development and maintenance of these is vital. For groups of invertebrates where a number of related species are involved, there may well be common elements to desirable rearing techniques. Concentration of long-term effort into some form of 'captive breeding institute', as Morton (1983) advocated for butterflies, may be viable. However, care must then be taken not to unknowingly mix up stocks from widely separated populations.

The techniques of practical aquaculture, though

designed primarily for contexts other than conservation, are important in demonstrating methodology that has potential for transfer to other taxa as part of conservation management. Therefore, it may not be necessary to start with a complete absence of knowledge relevant for a threatened taxon but, rather, hope to adapt techniques known to work well for related species, albeit in a different context. Techniques for the commercial rearing of freshwater crayfish, for example, have led both to detailed protocols for a number of species in various parts of the world and to a number of encouraging generalities relevant to conservation management through mass rearing. Thus, many species can be cultured, with excellent growth potential; food costs are low; they are resistant to most pathogens, and can tolerate poor water quality (Huner 1991). In Australia, methods developed for farming marron (*Cherax tenuimanus*) and the yabbie (*C. destructor*), exemplify possibilities available for other, non-commercial, species, and provide the economic background to indicate the feasibility of such technology transfer in small earthern ponds, where production levels of 1500–2800 kg/ha necessitate 'break-even' prices ranging from US$5.50 to US$10.50/kg (Morrissy *et al.* 1986, Stanford and Kuznecous 1988).

The aims of all such commercial operations include finding methods to increase survival and increase growth rate of the organisms. Most target species are native, but some are exotic: some freshwater prawns (*Macrobrachium*) have been shipped extensively outside their natural range. *Macrobrachium rosenbergii*, for example, is farmed in Hawaii and the Caribbean area, based on importation of original stock from Malaysia (J.H. Brown 1991).

The lessons learned by commercially motivated captive breeding programmes or husbandry, such as for edible crustaceans and molluscs, are clearly applicable more widely. Some of these have been for exotic species (such as the Pacific oyster, *Crassostrea gigas*, introduced to Australia in the late 1940s) (Dix 1990). Progress with raising this species led to an anticipated yield of 50 million oysters a year by 1990 from Tasmania. The same species has been cultivated extensively in Europe, and genetic manipulation has involved development of triploid and tetraploid oysters to increase yields (Héral and Deslous-Paoli 1990). The need for many such programmes has arisen directly because of the damaging effects of pollution on natural populations (see also the comment on clam mariculture, p. 144).

FARMING AND RANCHING

Two aspects of 'captive rearing', both important for conservation of certain invertebrates, differ somewhat in the derivation of the stock maintained:

1. 'Farming', implies entirely captive populations, with captive-bred parental stock maintained over a series of generations in enclosed conditions.

2. 'Ranching', implies harvesting or 'concentration' of field populations which are then exploited in a controlled way—perhaps by taking a proportion of early stages produced and rearing these under conditions where they are protected from major natural mortality, such as predators or parasites.

This dichotomy has been particularly evident for butterflies. One recommendation made in a recent conservation action plan for the swallowtail butterflies (New and Collins 1991) was that the feasibility of farming or ranching a number of the species should be investigated as a means to:

(1) alleviate collector pressure on wild populations by providing high-quality reared specimens for commercial sale; and

(2) build up stocks for release to the wild.

For butterflies, natural mortality of eggs and caterpillars caused by predators and parasitoid wasps can be very high, so that protection of ranched stock can lead to survival levels much higher than might occur otherwise. Until now, most 'butterfly farming' (including ranching) operations have been orientated primarily to satisfy commercial demand, and the conservation benefits have been largely secondary. However, with recent trends to expand the butterfly livestock market (p. 102, Collins 1987a), for which regular supplies are needed, there is a corresponding trend of increase in the number of butterfly farms in various parts of the tropics (New 1991a). Widespread potential for conservation enhancement is available through these.

This potential can undoubtedly be fostered effectively by the demonstrated positive local effects from some butterfly-rearing programmes, and the need to enhance and sustain these. One of the pioneer, and now best-known community-orientated insect ranching programmes, the Papua New Guinea-based operation, co-ordinated through the central Insect Farming and Trading Agency (at Bulolo, Morobe Province), was instigated in 1978 (National Research Council 1983; Collins and Morris 1985), and developed from a more local operation which started in 1974. An 'instruction manual' (Parsons 1978) on establishing a butterfly garden where stocks of birdwing butterflies, in particular, can be concentrated by provision of large numbers of foodplants (that is, by habitat enrichment), also provides information on specimen care. At present, more than 500 people in the country derive income directly from butterflies (Clark 1992), and this number is poised to increase considerably— to the extent where the butterfly industry will contribute substantially to the well-being of the rural sector. An average income from butterfly ranching in Papua New Guinea may be substantially larger than that derived from subsistence or traditional agriculture. By deriving cash from butterflies as a sustainable harvestable resource based on enrichment of areas already cleared of primary forests, the rate of future clearing of primary habitat may be reduced.

GENETICS

Understanding the genetic ramifications of captive breeding is important in attempting to address the long-term effects of inbreeding, or of the well-being of laboratory stocks commenced from relatively small founder populations. For invertebrates, the development of 'genetic protocols' is in its infancy, but these have strong practical relevance in assuring the fitness of individuals reared over many generations and of countering any likelihood of 'domestication', which could reduce their eventual capability or suitability for release in the wild. It may also be critical, in the interests of conserving genetic diversity within a species, to ensure that any releases are made in the same area from where the captive population derived (whenever that site is still suitable), and there are clear risks in releasing captive stock to augment an existing field population of probable different genetic constitution. Inadvertent mixing of genetic stocks may even mask evolutionary processes, such as speciation, through hybridization between distinct biological forms. 'Inbreeding depression' may occur in the laboratory at times, but the common reality with rare taxa is that there is little option other than to concentrate on obtaining as many offspring as possible from very few parent individuals. However desirable it may be in theory to conserve a wide genetic diversity, many of the invertebrates (and others) that command attention as conservation targets have already been reduced to very small populations, often to a single, small remnant.

Genetics in conservation

The place of genetics in modern conservation biology was reviewed thoughtfully by Templeton (1991), who exemplified four general areas of application. Although his emphasis was on vertebrates, the principles apply equally to the development of an appreciation of its relevance in invertebrate conservation. The areas are:

Conservation forensics
This can involve genetic techniques (predominantly DNA fingerprinting) to aid enforcement of laws on endangered species (Chapter 6) by providing individual identifications. This field now has greatest relevance to vertebrates, such as to identify sources of ivory, but may eventually have value in invertebrate trade, where it

Table 5.1 Policy factors that would be expected to minimize the loss of genetic diversity in a management programme for invertebrates (listed in descending order of priority) (from Brakefield 1991)

1. Maintain a large or substantial population size, and maximize the proportion of adults contributing to reproduction
2. Minimize the incidence of bottlenecks in population size
3. Minimize the duration of any bottlenecks
4. Maintain migration (= gene flow) between, and continuity of, any local populations
5. Maintain appropriate environmental heterogeneity

may be necessary to trace individuals to particular populations/sites.

Systematics

Accurate identification or recognition of taxa is clearly important, not only *per se* but also in assuring the integrity of techniques such as augmentation or translocation (p. 108) of natural populations, and to avoid the inadvertent mixing of distinct genotypes, even from different populations of the same species, at times (Ehrlich and Murphy 1987). The 'biological species concept' will continue to engender debate and, perhaps especially for invertebrates where morphological differences between taxa are often minute, any corroboration of population distinctiveness may be relevant to eventual management. As Templeton noted, 'before we can design a conservation programme, we need to know what it is we are trying to conserve'. Studies of homologous pieces of DNA, using such procedures as gene sequencing, to detect variations may aid in defining evolutionary relationships within a group of lineages (species, or populations).

Hybridization

Environmental disturbance, such as removal of barriers between previously separated populations by habitat change, or introduction of exotic species or of additional stocks of the same apparent species, can result in mixing of populations which were formerly discrete.

Genetic management

Templeton (1991) cited genetic husbandry, preservation of adaptive flexibility, and potential for genetic engineering as facets of genetic management aiding the primary goals of preserving genetic diversity and altering the genetic

environment of a species as little as possible, or (less ideally) adapting a species to necessary changes in the environment.

More specifically, several relevant genetical aspects of management for invertebrates were discussed by Brakefield (1991) (Table 5.1). Even with small populations—Brakefield cited intermittent bottlenecks of ten to a few tens of effective individuals—genetic variation is likely to be of less practical concern to conservation than the more usual anthropogenic causes of population extinction. It is therefore generally not necessary to foster maintenance of genetic variation as a management priority. This, in part, is because the prime aim is usually to assure maximum population size gain as rapidly as possible; more emphasis on genetic management is needed for a longer-term programme.

However, Brakefield (1991) cautioned that many species of conservation concern are subject to continuing loss of habitat and population fragmentation, so that the relationship between ecological success and genetic variability is usually not clear. Some insects that have suffered extreme bottlenecks of small population size can retain substantial ability to adapt to 'new' changes. Bottlenecks that extend over many generations are much more serious, and loss of viability for small populations (such as through loss of rare alleles) may well reduce the population's capability to withstand environmental stress in the future.

Long-term survival may *require* the maintenance of rare alleles, and adaptation to inbreeding alone is not a foolproof way to adapt to environmental change. For molluscs Selander and Ochman (1983) and Foltz *et al*. (1984) discussed several relevant examples: one is that many species of terrestrial slugs, capable of hermaphrodity, use outcrossing as their normal breeding system.

Lack of genetic variation may pre-dispose populations to extinction, but in outbreeding taxa extinction may be caused or accelerated by loss of alleles through genetic drift. Inbreeding depression may lead to lowered survival and a decrease in population size.

The evolutionary potential of small populations may depend on the potential for increasing genetic diversity. A general concern for small and isolated populations is that such potential is reduced markedly. For outbreeding species, two principal interacting factors can accelerate or precipitate extinction. These, noted above, are genetic drift and the actual population reduction through inbreeding depression.

1. Genetic drift involves the loss of alleles and changes in allele frequency that result randomly for each sexual reproduction. In essence, alleles in non-breeding individuals may decrease in frequency and may be lost, eventually, from the population. This normal genetic drift is enhanced in small populations because fewer individuals are available to reproduce. There is some suggestion that this trend results in an increase of deleterious alleles and decreased frequency of beneficial ones.

2. Inbreeding depression may result when normally outbreeding species are forced to inbreed, when the combined effects of (a) expression of deleterious recessive alleles, (b) loss of alleles caused by small population size, and (c) reduced representation of alternative alleles in the population lead to lower survival rates. Such considerations are highly relevant in the captive management of species.

Accepting that maintenance of genetic variability may be essential for long-term survival, the practical question then arises about how much conservation investment should be made in attempts to save very small populations, in which genetic drift may be high and where variation may decline rapidly and render their chances of persistence poor (Soulé 1980, Lande and Barrowclough 1987). This topic interacts with other aspects of 'importance' or 'priority'

(Chapter 3), but criticism has sometimes been levelled at highly expensive attempts to save regional populations (of butterflies, in particular), especially of taxa which are represented adequately elsewhere. This would include populations on the fringe of their geographical range, where the microclimate may be only marginally suitable for them to thrive.

Considering plant examples, Lesica and Allendorf (1992) indicated that small populations in stressful conditions might lose variability (heterogeneity) more slowly than in benign environments, although the contrary view has been expressed also (Frankel and Soulé 1981). Small populations, if they do retain greater than expected levels of variability, might also retain their capacity for adaptation for long periods. However, notwithstanding genetic constraints, which are generally poorly understood for rare invertebrates, small populations are likely to be at greater risk from random environmental fluctuations.

Many invertebrates have short life cycles and high reproductive rates, so that their genetic constitution has great opportunity to change—indeed, Morton (1991) commented that 'only a highly adapted form in the centre of a stable environment is likely to remain genetically constant'. Considering insects, Morton noted several important ramifications of this condition in relation to captive breeding and *ex situ* conservation, namely:

1. Populations are not interchangeable. Local populations of the same species may differ considerably, for example in the amount of phenotypic variability, preferred larval foodplants, and tolerance to local climatic regimes.

2. Populations are not genetically static. The frequency of genetic change means that local 'forms' lack any 'eternal value' despite their undoubted taxonomic or ecological interest (Berry 1972).

3. Species and local forms cannot be preserved indefinitely. This necessitates an approach towards conserving the future evolution of a species, rather than simply the *status quo*.

Genetics and captive breeding

In addition to questioning the 'conventional wisdom' of conserving such taxa, this also emphasizes that habitat conservation on a sufficiently large scale to preserve the full range of intraspecific genetic diversity is unlikely to be feasible. Morton (1991), therefore, queried whether captive breeding as a conservation strategy may lead to production of individuals with a generalized gene pool that is not adapted to *any* natural environment.

Long-term captive breeding is akin to domestication, with attendant likely loss of traits that favour survival in the wild. Conserving variability may be difficult in many instances where only very small founder populations are available. Once a captive population has undergone a few generations, strong selection pressures are likely to occur, especially if the maximum sustainable population size (that is, the largest that can be housed or maintained adequately) has been reached, and these adapt the species to the captive environment. Morton noted three possible practical options that might help counter this by maintaining population size at lower levels:

1. Restricting natality by minimizing juvenile mortality. Selection thereby occurs only in determining which adults are used for breeding, and this may have the severe disadvantage of minimizing the effective population.

2. Encouraging natality but increasing juvenile mortality. Allow all adults to mate and lay, then select sample of eggs randomly for rearing. Alternatively, allow all the offspring to develop, but under conditions where high mortality may occur, so that survival fitness is selected for; however, the strong selection pressures in this scenario may themselves be disadvantageous.

3. A mixed strategy, seeking an intermediate path which might result in retaining more variation than either of the two previous options.

A number of guidelines for insectary management (see also Lees 1989) suggested by Morton

(1991) merit serious consideration in the interests of maintaining biologically representative captive stocks of many invertebrate taxa:

1. Start with large founding populations, ideally from large numbers of immature stages.

2. While founding the captive populations, attempt to minimize mortality and maximize reproduction, to build up the population rapidly.

3. Then, reduce the population to a 'manageable size' by imposing selection pressures (such as reasonable levels of competition, or fluctuating climatic regimes) on the immature stages.

4. Use adequately large mating cages to encourage mate-location and courtship behaviour. For moths, improve airflows to remove pheromone concentrations, and the like.

5. If possible, select females for appropriate host-finding and oviposition behaviour.

6. Maintain separate lines, and cross these systematically. Monitor the performance of parental and hybrid lines and do not discard lines except those that consistently perform badly.

7. Monitor changes in gene frequency over time, in each captive line, for example by measuring frequencies of biochemical markers. This approach may be useful in correcting deficiencies in particular lines.

This general 'genetic protocol' poses important questions of logistic capability and the need to change more established practices in order to conserve the greatest variation and, hence, potential for successful release to the wild. It is relevant also in contexts such as the mass-rearing of biological control agents and, thus, in assessing the characteristics of what may be exotic species being introduced deliberately to a new environment. However, it may be restricted in practice by a small founding population of a rare species.

The preliminary genetic protocol for the red-kneed bird-eating spider was noted on p. 101.

One of the few other cases where such a protocol has been derived is noted briefly here, to illustrate the emerging awareness of the need for this aspect of captive management of invertebrates.

The genetic protocol developed for *Partula* snails (Tonge and Bloxam 1991) has three main steps, involving the need for:

(1) six different cultures of each taxon to be established;

(2) each culture containing two sets ('inbred' and 'crossed') of 25 individuals, housed in five boxes each with five adult snails; and

(3) the maintenance of distinct non-overlapping generations.

The rationale underlying this is that each line should retain some genetic variability, with the inbred lines less susceptible to selection pressure because of pressures of inbreeding and genetic drift. These will indeed lose genetic variability, but will be heterogeneous between lines and will suffer high extinction rates because of inbreeding. The crossed lines should have a higher expectation of survival, through having greater heterogeneity.

It is very difficult to predict fully the genetic effects of captive breeding, and there is usually little information on the representativeness or heterogeneity of the founder stock, even if the number of individuals used is known. The test of fitness on release from captivity has not been quantified well, and inferences on the reasons for success or failure of releases are often ambiguous.

For the rare carabid beetle *Chrysocarabus olympiae*, Malausa and Drescher (1991) noted that during 13 generations in captivity (as of 1990, for two founder populations of around 10 individuals each, collected in 1975 and 1979 respectively), no morphological malformations, decrease in size, or signs of 'degeneration' were found. Reduction in individual size is a common manifestation of reduced fitness in reared Lepidoptera and other insects. Fluctuating asymmetry (p. 29) may prove to be a useful index for genetic 'quality control' of captive stocks of invertebrates because of its sensitivity in revealing developmental stresses before more obvious changes in life-history parameters are manifested. It has been recommended as a quality control indicator for insect mass-rearing programmes for biological control (Clarke and McKenzie 1992), and as an indicator of inbreeding stress (Clarke *et al*. 1986).

For most invertebrates of direct conservation concern, little (if any) genetic information is available from natural populations as a basis for initiation of management programmes involving captive rearing. The coconut crab (p. 163), for an example of a rare exception to this, has been studied over part of the Pacific. No genetic differences were found between crabs from various islands of Vanuatu, or between crabs from these areas and other Pacific island groups, but there were distinct differences between 'Pacific Ocean' crabs and ones from Christmas Island (Indian Ocean). Such differences pose further questions for assuring the integrity of captive management—such as whether brood stock movement between different islands should be prevented (that is, is it compatible with maintaining genetic variability?), and the avenues of recruitment to the collective Vanuatu population (Brown and Fielder 1991).

The relevance of population genetics to insect introductions was summarized by Smith and Holloway (1989), who emphasized that practical opportunities to study this have been limited almost entirely to deliberate introductions, mainly of biological control agents. Introducing 'new blood' to augment small existing populations can inadvertently cause problems of reduced fitness, but these must be balanced against the likelihood of extinction which might occur without this step.

Changes in some biological control agents following their introduction to new regions (Murray 1985) are also relevant: although the immediate context is different in emphasis from conservation biology, the principles converge strongly because the practical desire is to assure the success of a released organism. In appraising strategies for introducing control agents for plant weeds, Murray suggested two strategies based on genetic parameters, and which differed according to the nature of the receiving environment.

Table 5.2 Definitions of terms involved in releasing invertebrates to the wild (JCCBI 1986)

Term	Definition
Re-establishment	Deliberate release and encouragement in an area where the invertebrate formerly occurred but is now extinct
Introduction	Attempt to establish in an area where the species is not known to occur, or to have occurred
Re-introduction	Attempt to establish species in an area to which it has been previously introduced unsuccessfully
Reinforcement	Attempt to increase the population size by releasing additional individuals into the population
Translocation	Transfer of individuals from an endangered site to a protected or neutral one
Establishment	A neutral term: any attempt to intentionally and artificially increase numbers by transfer of individuals

1. If the new environment resembled that of the parental (source) population, it may be advisable to attempt to represent genetic variability and avoid any inbreeding depression. This could be achieved by:
   (a) mixing individuals from different sources;
   (b) maintaining them as a large stock without adapting them to artificial rearing regimes; and
   (c) making a few releases, each of large numbers of individuals.

2. If the new environment differs substantially from the parental one, the preferable strategy would be to:
   (a) take a small number of stock individuals from one source population;
   (b) increase the captive population rapidly in the new environment; and
   (c) *either* split it into many separate lines soon after release *or* establish many separate lines from small initial samples. Most of these would become extinct after release, but the survivors would be adapted well to the new environment.

In view of the likely increase of transgenic agents against pests, Hoy (1992) has called for the development of protocols for evaluating the risks of releasing these, and noted that ecologists and conservationists should be involved in formulating guidelines for use of this novel class of exotic organisms.

TRANSLOCATION AND RE-INTRODUCTION

Translocation and re-introduction are the most common roles for released captive stock (Table 5.2). Both are forms of 'introduction', and were defined by the Joint Committee for the Conservation of British Insects (JCCBI) as part of a suite of related terms which, in practice, tend to be used interchangeably. Translocation refers to the removal of stock from one place to another and need not, necessarily, incorporate any captive phase. Likewise, stock may be re-introduced to an area where the species was present formerly, whereas translocation usually implies introduction to a 'new' site—perhaps to extend a species' range, or perhaps to a previously unoccupied site within the species' historical distribution. Some introductions, of these and other sorts, are controversial, and few biologists would condone haphazard or poorly planned 'random' releases of taxa of conservation significance; or, indeed, of exotic species released deliberately to enhance a local fauna. The JCCBI Code of Practice (Table 5.3) discusses some of the widespread concerns that can arise. The importance of translocation is increasing rapidly.

Conant (1988) discussed the general principle of translocation to establish populations of endangered species outside the normal range. She pointed out that changes to the original

Table 5.3 Summary of the main recommendations from *Insect re-establishment – a code of conservation practice*) (JCCBI 1986)

1. Consult widely before deciding to attempt any re-establishment
2. Every re-establishment should have a clear objective
3. The ecology of the species to be re-established should be known
4. Permission should be obtained to use both the receiving site and the source of material for re-establishment
5. The receiving site should be appropriately managed
6. Specific parasites should be included in re-establishment
7. The numbers of insects released should be large enough to secure re-establishment
8. Details of the release should be meticulously recorded
9. The success of re-establishment should be continually assessed and adequately recorded
10. All re-establishment should be reported (in this code, to the Biological Records Centre and the JCCBI)

habitat may have been the predominant initial endangering factors, and that the species' original sites may have been altered to the extent that re-introduced populations might not survive. Two kinds of danger, one 'more practical' and the other 'more fundamental', were foreseen. The first derives from the translocated species being, in essence, exotic in its new area and that it might do environmental harm there. Conant cited the instance of the Nihoa millerbird (*Acrocephalus familiaris kingi*), endemic to Nihoa (Hawaii) and proposed for introduction to the nearby Necker Island. Necker supports at least 15 species of endemic terrestrial arthropods and *no* avian insectivores, and could probably sustain only about seven pairs of millerbirds. The proposal was not implemented. The second is the possibility of evolutionary change: classic 'island isolation' differentiation, sometimes fostered through captive breeding over several generations, which could change the genetic nature of the translocated population. In this context, translocation is a conservation-orientated equivalent to introduction of an exotic species for any other purpose (such as biological control).

There is thus strong feeling that any attempt at translocation or, more generally, at introduction or re-introduction, *must* be responsible and documented soundly rather than undertaken casually , however good the intention. The JCCBI code (1986) is of wide relevance, although designed particularly for insects. The general steps involved, for either species-orientated re-establishments (mainly involving those species which are endangered or vulnerable) or site-orientated operations (by which the 'wildlife'

or amenity value of a site is increased by providing species which were at some time present there) are:

(1) appraisal of the site's features, including any aspect of management or change which may be needed, and the feasibility of undertaking this;

(2) preparation of the site to ensure that any released species can indeed survive, in relation to its known resources requirements;

(3) introduction of stock without depleting the source population, and of a kind adapted to the conditions of the 'new' site; and

(4) monitoring the re-establishment attempt to determine success or failure, and the reasons for this.

Each of these steps has a number of different facets, and the major recommendations are noted in Table 5.3.

As well as the need to document any transfers of invertebrates, JCCBI (1986) suggested that advice should be sought to determine whether any 'controls' or restrictions are in place or are needed for particular cases. Conservation is only one of the reasons for translocation, and the wider rationale is exemplified by Horwitz's (1990*a,b*) survey of the reasons for moving freshwater crayfish, namely:

(1) for scientific interest and research, and/or exploration of aquaculture potential;

(2) provision of live animals for the restaurant trade;

(3) provision of live animals for the aquarium trade;

(4) full aquaculture exploitation; and

(5) stocking of farm dams for recreation and harvesting for household consumption.

Any of these may involve subsequent release of stock to nearby waterways in order to dispose of it (Horwitz 1990a), so that non-endemic or non-local species may well be given the opportunity to establish sustainable populations.

The concerns are essentially those of introducing an exotic species or genotype, and parallel those of translocating rare species to new areas. They are likely to increase in importance with increasing frequency and need of translocations, and protocols to minimize any unwanted effects are needed urgently. For Australian crayfish, Horwitz (1990a) recommended a combined regulatory and educational programme to address five main conservation-related parameters:

(1) preventing release of aquarium or farm crayfish into waterways;

(2) preventing importation of species not native to Australia (this maintains current practice, in the face of industry pressures to relax it);

(3) ensure that stock translocations on all scales are declared and recorded;

(4) ensure that disease-free 'status certificates' and quarantine measures accompany all translocations, and that national standards are reviewed to ensure capability for this; and

(5) ensure that people certifying stocks are experienced in invertebrate pathology, and do not have vested interests in the crayfish industry.

The complexity of these recommendations illustrates well the range of conservation concerns, and the need for both scientific and political responsibility: see Chapter 6. The genetic distinctiveness of particular populations may prevent (or, at least, militate against) movement of stocks of the same species between different localities.

Horwitz emphasized (following from examples discussed by Holdich 1987 and Lindqvist 1987) that introductions or translocations of crayfish for purposes such as population restoration or enhancement must be based on research-founded management strategies. The potential for hybridization between closely related taxa may lead to loss of distinctive characteristics of both populations (p. 59) with little practical chance of re-establishing these in their pristine condition.

Despite the need for translocation or introduction in particular conservation programmes, this need for responsibility must be emphasized. The Xerces Society (p. 126) put forward their viewpoint on translocation of Lepidoptera as follows: 'the introduction of live individuals as a conservation measure is extreme and potentially dangerous' (Pyle 1976). It should be undertaken, they believe, *only* in the circumstances of population extinction, within the historical range of the species, or with introduced material of identical genome to the remnant population.

The latter can be very difficult to assure, especially after a period of captive breeding. Several relevant parameters of successful translocations of insects, discussed by Lees (1989), are:

1. Sustainable habitat for re-introduction must be available: if it is not, the risk to captive stock should be reduced by dividing it into subpopulations in several insectaries (the approach exemplified by the *Partula* programme, p. 100).

2. The captive populations should have been founded from stocks which should not have been subject to population bottlenecks. In some cases this is impossible, because only very small founder populations may be available for endangered species.

3. The captive stock should not have reduced 'fitness'.

4. The site for re-introduction should be capable of being protected or supervised.

5. Adequate funding must be available for captive breeding programmes and for habitat maintenance.

6. There should be no 'political impediments' or lack of funding to obtain any necessary permits for importation or release, if these are needed.

FURTHER READING

Collins, N. M. (ed.) (1990). *The management and welfare of invertebrates in captivity*. National Federation of Zoological Gardens of Great Britain and Ireland, London.

Nash, C. E. (ed.) (1991). *Production of aquatic animals. Crustaceans, molluscs, amphibians, and reptiles*. Elsevier, Amsterdam.

Olney, P. J. S. and Ellis, P. (ed.) (1991). *1990 International zoo yearbook*. Zoological Society of London, London.

6 REGULATION, LEGISLATION, AND ASSESSMENT OF STATUS

NEED FOR FORMAL CONTROL

The last three chapters have stressed some of the biological components of invertebrate conservation. In them, repeated allusions were made to the other major component of most conservation programmes: some form of regulation or 'protective legislation', the need for and implementation of which may arise in various contexts. For example, there may be a need to control the degree of exploitation of a natural population by people, and this is the basis for most 'protective legislation' and less formal regulations. Effective quarantine may be needed to prevent ingress of exotic taxa, or steps may need to be taken to attempt to eradicate them once they have arrived. Any such regulatory steps need to enforced and to be based on very sound information on the status and distribution of the taxa at whose protection they are directed. Some form of regulation is a common and important part of countering virtually any of the threatening processes noted in Chapter 3, but it is beyond the scope of this book to discuss, for example, the complex array of anti-pollution laws that exist in many parts of the world. I want, rather, to indicate the complexity of, and need for, regulatory aspects of invertebrate protection, to indicate some shortcomings of these, and to outline their use and application in practical conservation of invertebrate species and populations, and in the allied field of assuring sustainability of commercially harvested taxa.

The need for any regulation or 'formal protection' arises from consideration of the impacts of

perceived threats, and their assessment can be very complex. Suites of threats may occur separately, or combined in various ways. In a situation of people collecting coastal marine invertebrates, for example, pilot studies in New South Wales, Australia (Underwood and Kennelly 1990) led to recognition that variables such as level of exposure (tide times), site accessibility, and weather, influencing short-term exploitation, and others, such as differences in human activity levels between school holidays and school terms, and weekdays and weekends, need to be included in any long-term assessment of impact. People tend to glean on the coast more on sunny days than in poor weather, and are more numerous over holiday periods than others. They may also be ignorant of the need for protection, or of the existence of prohibitions. In Victoria, Australia, collection of most intertidal molluscs is prohibited by the Shellfish Protection Zone. Many people still harvest food or bait species, although prominent notices displaying the attendant fines are sometimes present about every 500 m along the shore (Keough *et al.* 1993).

Among the effects on invertebrates of the rock platforms in the New South Wales study were:

1. Most individuals of taxa used for bait or food were removed from a habitat patch, except for any of undesirable minimum size levels, set by preference for eating or difficulty of harvesting.

2. Therefore, organisms were left intact at some sites but all were removed from others,

increasing the 'patchiness' of the species involved.

3. There was a tendency for species removed to be the larger ones, possibly equating to 'keystone species' (p. 23) of the coastal habitat, with important effects on other species present.

Commercial species

This scenario reflects the localized effects of prime concern when attempting to safeguard or conserve invertebrates on a particular site and where correspondingly local controls may be needed, even if they are difficult to enforce. The use of formal regulations is also widespread in commercial fisheries, and the range of options developed there also indicates the approaches which might, in due course, be applied in conservation programmes. Few exploited invertebrates in marine ecosystems are currently threatened severely, but some may be vulnerable (Jamieson 1993) due to the ease with which they are captured, exceptional market demands, their limited distributions, or long life cycles associated with a low annual recruitment rate. Rounsefell (1975) pointed out that there are two categories of regulation which either change the amount or kind of human 'predation' on such populations. Either the total catch can be reduced or limited, or (/and) certain portions of the population can be protected. His comments, particularly relevant to marine fisheries, are of very broad application and relevant (with appropriate modifications) to other environments and taxa.

The first of these approaches comprises three main strategies:

1. Lowering the *efficiency* of an individual fishing unit is relatively easy to achieve by imposing restrictions on the size or kind of vessel or gear, closing areas to particular kinds of gear, placing bag-limits or quotas, and restricting fishing time or opportunity.

2. The *number* of fishing units can be restricted, a strategy referred to as the 'limited entry' system, so that the catch may be reduced without reducing the efficiency of the gear.

This system can become very complicated to execute and may be unpopular politically as livelihoods may be affected directly through exclusion from a resource which is regarded by many people as public property.

3. Limiting the amount that can be taken in any season (or any given time period) by imposition of a quota. This presupposes adequate biological knowledge to ensure that the quota is realistic in relation to population sustainability. This knowledge is most commonly not available, and 'blind quotas' are not uncommon, on the basis of 'erring' on the side of safety.

The other objective (protecting portions of the population) has a range of desirable effects which can be brought about in up to five main ways (Rounsefell 1975):

1. The 'savings gear' method involves modifying gear, such as having minimum mesh sizes in nets or minimum spacing of escape ports in crustacean pots, so that undersized individuals are not captured or can escape easily.

2. Closure of designated areas to fishing is a measure applied to breeding refuges such as estuaries where shrimp concentrate during their main growth season. It can be applied to polluted areas of water, as another example, where it may gain sympathy in being seen to safeguard human health.

3. Closed seasons are also used to protect breeding stocks, or other critical growth stages or size classes.

4. Sales restrictions, usually based on size, are sometimes used—a strategy reflected in prohibition of catching particular size classes (crustaceans, molluscs, sponges—though the last are often 'trimmed' before sale). This leads to the next category.

5. Individuals in particular phases are protected. Examples are moulting crabs of some species and some lobsters and crayfish when 'berried' (that is, carrying eggs). There is often also an element of pragmatism in such practices, and this may replace

regulation, especially for non-commercial purposes. Collectors of butterflies, for example, want 'perfect specimens' and there is little value in capturing worn or ragged individuals. This point is included in several 'codes' for collectors; however, these are usually 'advisory' rather than regulatory, as they are not enforceable legally.

Restrictions on exploitation may thus be driven by market opportunity: the higher the demand for a commercially desirable species, the greater the effort warranted to exploit the species in greater numbers or more efficiently. Without regulation, in such situations the species is 'safe' only when it becomes too rare to be 'worth the effort' of obtaining it: that is, assuming that it has not been reduced to non-viable population levels by that time. This economic restrictive parameter does not apply to all invertebrates, because avid collectors will go to extraordinary lengths to obtain rare species of shells, beetles, or butterflies, for example, and continue to hunt fervently to capture them. It is more for this kind of situation that much 'protective legislation' has been devised, and is much harder to render effective. Many commercial-crop species restrictions are accepted readily by fisheries industries, simply because they are seen to help in sustaining the long-term interests of the industries by protecting stocks from overexploitation, and ensure maintenance of adequate (or, at least, minimal) breeding stocks. Poaching and transgression of bag—or size—limits, still occur, but the penalties tend to be severe (including confiscation of gear and, even, of boats) and are a major general deterrent to this.

PROTECTIVE LEGISLATION AND CONSERVATION OF SPECIES

Legislation aimed at preventing the collection of species can broadly address the integrity of natural communities, or focus on taxa of particular conservation interest.

In the broader context, collection of invertebrates (and other biota) is commonly prohibited, without a permit to do so, in high-profile reserve areas such as national parks, in many countries. The need for permits must usually be justified soundly by people wishing to collect there. Restrictions of this sort thereby function as 'umbrella legislation', with the purpose of sustaining an area's natural biological integrity. Commonly, though, they are very difficult to enforce for invertebrates, and have far greater emphasis on, and priority for, vertebrate animals. Many 'collectable' invertebrates are small and can be obtained and carried easily without recourse to unwieldy or conspicuous equipment.

The more specific context of protective legislation is that which is applied to individual nominated taxa perceived to be vulnerable or endangered, or otherwise likely to come to harm if they are collected. A total ban on taking individuals of those species is usually imposed, sometimes with accompanying provision for very low numbers to be captured by specific licence or royalty payment. This kind of legislation has tended to result in polarized opinion among invertebrate conservation practitioners—depending, in part, on the effectiveness of an individual piece of legislation. Thus:

1. Supporters of the principle advocate progressive selection and legal protection of more and more threatened invertebrates for which it can be demonstrated that vulnerability might be increased by uncritically taking them from the wild.

2. Opponents query whether the latter situation occurs; they emphasize that habitat protection is far more important as a conservation measure, and suggest that designation of a species as 'protected' enhances its commercial status. They also raise other significant points noted below.

In short, this aspect of invertebrate conservation is controversial as to its practical effects. The intention is, usually, sound: to prevent overcollecting and possible resultant endangerment to the taxon concerned. However, in the past, a number of species have been listed for protection as a result of the zeal of individual scientists/conservationists, rather than from any sound knowledge of the species' biology, so that listing can be open to easy discreditation in some instances. It has also been extended, sometimes, far beyond the protection of notable or deserving species. Thus, in some European countries (Collins 1987b) prohibition of collecting extends to 'all butterflies' or 'all butterflies except white-winged pierids', or some similar blanket grouping. The most extensive case appears to be for the Lagintal area of Switzerland, where all Lepidoptera are protected and it is illegal to carry a butterfly net; this is under the guise of protecting the local endemic satyrine butterfly *Erebia christi*. Such extremism, even when fostered sincerely by concerned conservationists, can lead to ridicule and resentment from the very people who could do most to help the species concerned, namely knowledgeable collectors and naturalists. It is not unusual, under such extensive listing, for very common species with no perceived threats to be totally protected legally, and with substantial penalties able to be imposed for killing or capturing them. People find this situation difficult to understand, and it can deter 'hobby interest' (Chapter 8) substantially and bring protective legislation into disrepute.

Part of the 'rationale' for such broadening has been that any protective legislation has to be monitored and enforced in the field by non-expert personnel. They may not be able to recognize or differentiate similar-looking taxa, only some of which are legitimate targets. Many closely related butterflies or beetles, for example, are distinguishable only by specialists, perhaps only after dissection to examine genitalia or other intricate structural details. Broadening of legislative protection to cover all possible taxa likely to be confused with those of 'real' concern is one way to obviate this problem. A similar approach to 'look-alike' species may be adopted in listing species for which trade is restricted. Thus, all birdwing butterflies are listed on the Convention on International Trade in Endangered Species (CITES, p. 117) even though some species are common and are under no threat.

However, this relatively rare advantage is outweighed commonly by the suspicion and alienation that result. Other criticisms have been levelled at legislation of this kind:

1. It is extremely difficult to police, especially in remote areas with low staffing levels of rangers who may not have specialized knowledge or access to museum collections, or voucher material of the species they need to monitor.

2. Members of the public generally assume that prohibition of specimen-taking is accompanied by provision for habitat protection and ecological study of the species involved, to facilitate practical conservation and informed management. Usually, neither adjunct occurs, and there are instances of 'a species' being protected legislatively while its critical or restricted habitat continues to be alienated or destroyed. Legal protection *per se* is thus often regarded as 'toothless'—it can even be misleading, because people see it as positive whereas it may not be of any practical value for the taxa, and can even provide 'camouflage' for destroying it through habitat loss.

3. For commercially desirable rare species, highlighting them by singling out for individually designated protection can serve to increase their public profile, and lead to increased levels of illegal exploitation to satisfy an artificially increased black market demand, and commercial value.

4. The extensive bureaucracy associated with obtaining permits to collect or study a protected species may deter the very research that is needed to understand and conserve the species. Especially to amateur enthusiasts, who have played major and important roles in documenting invertebrate biology and distribution, such procedures are very daunting. There are instances (for example for butterflies) (New 1991a) where legal protection has led to diminished interest in the species, because people have opted to study or work on taxa that are more easily accessible to study.

In general, the most relevant and reasoned legislation to protect invertebrates is in the temperate regions, where species-focusing has long been important in conservation, and where there is provision for informed debate and review of particular cases and taxa nominated. Nominations to list species under the United States Endangered Species Act (ESA), or the

United Kingdom Wildlife and Countryside Act, for example, are not frivolous, and substantial scientific rationale and provision for management are an integral component of nomination. For ESA, for example, the two categories of listed species are those 'endangered' ('in danger of extinction throughout all or a significant portion of its range') and 'threatened' ('likely to become endangered within the forseeable future'). Recovery plans identify specific tasks needed so that a species can be downgraded from endangered to threatened or delisted. Likewise, in Australia the Victorian Flora and Fauna Guarantee Act necessitates the review by an independent scientific committee of the status of any species nominated, and preparation of a management plan should nomination be seen to be justified. This Act also provides for protection of rare invertebrate associations or communities. The legal entity known as 'Butterfly Community No.1', for example, is a particular association of species that includes two extremely rare hill-topping Lycaenidae, and is known at present to occur only on a small, isolated hill (Mt. Piper, near Broadford, north of Melbourne). Projected management includes protection of the mountain *and* of its butterfly fauna. It can also control 'threatening processes' and a number of inputs to running waters in the State, and removal of fallen branches from them, are illegal—though extremely difficult to monitor or control.

Such legislations can be seen to be responsible, but are logistically intensive to render effective. In some other parts of the world, such provisions do not exist, and projected regulations may not be declared formally or implemented. The Indian Wildlife Protection Act lists more than 450 butterfly taxa as protected, with substantial penalties for their capture. Many of the species or subspecies could be differentiated or recognized only by a few specialists in the groups concerned, and the rationale for their listing is not always clear.

It is sometimes possible to combine 'listing' of species for protection with a strong positive thrust for practical conservation, especially if the taxa can be considered as 'flagship species' (p. 25). The Papua New Guinea government nominated seven spectacular species of birdwing butterflies (Papilionidae) as 'national butterflies' in 1976, with the aim of protecting these rare and commercially desirable taxa. This was accompanied by the establishment of the national Insect Farming and Trading Agency (p. 20), the only legal avenue through which insects can be exported from the country.

Amongst the highest level of legal protection accorded to any invertebrate is the designation of a localized Queensland lycaenid, Illidge's ant-blue (*Acrodipsas illidgei*) as 'permanently protected fauna' in the state (Samson 1993). This status has been accorded otherwise only to high-profile vertebrates and means that, as well as collection of individuals being prohibited, permits are needed to hold specimens in collections or to transfer specimens between addresses or out of Queensland.

In addition to listing for specific local legislative protection in such ways, a number of invertebrates are included in various international conservation conventions, such as the Bern Convention. The aim of the Bern Convention (the Convention on the Conservation of European Wildlife and Natural Habitats, which came into force in 1982) is to 'conserve wild flora and fauna and their natural habitats' and 'give particular emphasis to endangered and vulnerable species'. By 1991, 51 insect species were listed, together with some other taxa, including 22 molluscs, for a total 81 species of invertebrates, responding to nominations sought by the World Conservation Union. Criteria for nomination were that the species should be under serious threat in Europe as a whole (not, necessarily, everywhere), be reasonably easy to identify, and have a predominantly European distribution. The recent European Economic Community Habitats Directive (European Community 1992) has a specific objective to establish a network of protected areas to ensure effective implementation of the Bern Convention, and several annexes to the document include invertebrates to be protected. Annex 2 lists species whose conservation requires designation of special areas of conservation (which are not defined). Invertebrates listed are one crustacean, 36 insects, and 22 molluscs. Annex

Table 6.1 Invertebrates listed in appendices of the Convention on International Trade in Endangered Species (CITES) (continued on p.118)

Appendix 1
  Insecta: Lepidoptera
    Ornithoptera alexandrae, Queen Alexandra's birdwing
    Papilio chikae, Luzon swallowtail
    P. homerus, Homerus swallowtail
    P. hospiton, Corsican swallowtail

  Mollusca: Unionoidea
    Conradilla caelata, birdwing pearly mussel
    Dromus dromas, dromedary pearly mussel
    Epioblasma curtisi, Curtis' pearly mussel
    E. florentina, yellow-blossom pearly mussel
    E. samsonia, Sampson's pearly mussel
    E. sulcata perobliqua, white catspaw mussel
    E. torulosa gubernaculum, green-blossom pearly mussel
    E. torulosa torulosa, tubercled-blossom pearly mussel
    E. turgidula, turgid-blossom pearly mussel
    E. walkeri, brown-blossom pearly mussel
    Fusconaia cuneolus, fine-rayed pigtoe pearly mussel
    F. edgarian, shiny pigtoe pearly mussel
    Lampsilis higginsi, Higgins' eye pearly mussel
    L. orbiculata orbiculata, pink mucket pearly mussel
    L. satura, plain pocketbook pearly mussel
    L. virescens, Alabama lamp pearly mussel
    Plethobasus cicatricosus, white warty-back pearly mussel
    P. cooperianus, orange-footed pimpleback mussel
    Pleurobema plenum, rough pigtoe pearly mussel
    Potamilus capax, fat pocketbook pearly mussel
    Quadrula intermedia, Cumberland monkey-face pearly mussel
    Q. sparsa, Appalachian monkey-face pearly mussel
    Tosolasma cylindrella, pale lilliput pearly mussel
    Unio nickliniana, Nicklin's pearly mussel
    U. tampicoensis toecomatensis, Tampico pearly mussel
    Villosa trabalis, Cumberland bean pearly mussel

  Mollusca: Stylommatophora
    Achatinella spp.

4 (species of community interest in need of strict protection) includes a longer list (46 insects, one spider, 23 molluscs, and one echinoderm), and Annex 5 (species whose taking in the wild and exploitation may be subject to management measures) incorporate one coelenterate, four molluscs, one annelid, four crustaceans, and one insect. A diversity of purposes is therefore incorporated in this document, but the Directive—in common with many similar conservation policy documents—seems unlikely to be implemented effectively because of paucity of funding. Others include the Convention on International Trade in Endangered Species of Wild Fauna and Flora (CITES, operative from 1973), which includes invertebrates in two appendices (Table 6.1). Appendix 1 lists species for which trade is prohibited entirely unless special export and import permits are granted under highly exceptional circumstances: these species are considered 'endangered' (p. 169), and no taking of further specimens can be countenanced at present without very clear justification. Appendix 2 lists species for which concern exists that trade could be a factor increasing vulnerability, and for which the extent of trade should be monitored through permit systems. Trade in any invertebrate species is extremely difficult to monitor

Table 6.1 *cont.*

Appendix 2
  Insecta: Lepidoptera
     Bhutanitis spp., Bhutanitis swallowtails
     Ornithoptera spp. (except *O. alexandrae*), birdwing butterflies
     Parnassius apollo, mountain apollo
     Teinopalpus spp., Kaiser-I-Hinds
     Trogonoptera spp., birdwing butterflies
     Troides spp., birdwing butterflies

  Arachnida: Araneae
     Brachypelma smithi, Mexican red-kneed tarantula

  Annelida: Hirudinoidea
     Hirudo medicinalis, medicinal leech

  Mollusca: Unionoida
     Cyprogenia abertj, edible pearly mussel
     Epioblasma torulosa rangiana, tan-blossom pearly mussel
     Fusconaia subrotunda, long solid mussel
     Lampsilis brevicula, ozark lamp pearly mussel
     Lexingtonia dolabelloides, slab-sided pearly mussel
     Pleurobema clava, club pearly mussel

  Mollusca: Stylommatophora
     Papuastyla pulcherrima, Manus green tree snail
     Paryphanta spp. (all New Zealand species), amber snails

  Anthozoa
     Antipatharia spp., black corals
     Pocillopora spp., brown stem cluster corals
     Seriatopora spp., birds nest corals
     Stylophora spp., cauliflower corals
     Acropora spp., branch corals
     Pavona spp., cactus corals
     Fungia spp., mushroom corals
     Halomitra spp.
     Polyphyllia spp., feather corals
     Favia spp., brain corals
     Platygyra spp., brain corals
     Merulina spp., merulina corals
     Lobophyllia spp., brain root corals
     Pectinia spp., lettuce corals
     Euphyllia spp., brain trumpet corals

  Hydrozoa
     Millepora spp., Wello fire corals

  Alyconaria
     Heliopora spp., blue corals
     Tubipora spp., organpipe corals

effectively, but highly publicised prosecutions for transgression may be a partial deterrent, and the declared figures on volumes of trade in particular taxa may be of some value as a guide to the extent of legal trade demands.

The monarch butterfly (*Danaus plexippus*, Nymphalidae) is included in the Convention for Protection of Migratory Species of Wild Animals (the Bonn Convention, operative from 1979), because of its spectacular long-distance migrations in North America, and the vulnerability of its traditional overwintering sites in California and Mexico (Brower and Malcolm 1989); the vast aggregations of overwintering adults were heralded as a 'threatened phenomenon' by Wells *et al.* (1983). Some other conventions do not yet include invertebrates. The 'Ramsar Convention' (Convention on Wetlands of International Importance, operative from 1975) for protection of wetlands, for example, does not yet cite them specifically, although there is a clear case for their importance as critical components of the ecosystems targeted for attention. It is highly likely that consideration will be given to fresh-water invertebrates under Ramsar in the near future, because of the wide importance of these habitats to animals in addition to birds, for which the convention was primarily designed. For the present, water-birds are serving as umbrella taxa for invertebrates.

For much of the general 'wildlife legislation' in force for many countries, it is by no means clear what the role for invertebrates may be, because terms such as 'wildlife' or 'fauna' are not always defined clearly in the relevant wording. Even when invertebrates are designated specifically, many practitioners equate these terms with 'vertebrates' alone, sometimes only to warm-blooded vertebrates. The profile of invertebrates in any such generalized regulation is most commonly low, and they are frequently overlooked completely.

Legislation that has as its objective the conservation of particular species has two major responsibilities, in addition to being capable of effective implementation.

(1) to ensure that species listed do indeed merit such listing; and

(2) to ensure that collecting prohibitions or restrictions, such as closed seasons or 'bag-limits', are based on the best information available and are likely to be an effective part of management to sustain the species involved.

The first of these has the implication that listed species should have priority over non-listed species in allocation of scarce resources available for practical conservation. The second implies the need for quantitative information on population status and the factors that affect it, and is idealistic for most species. Clearly, the criteria and justification for selecting any invertebrate from the myriad taxa available for such priority is a critical process, and imposition of restrictions should, wherever possible, have the 'sympathy' of the users of any commodity species involved, and be seen to protect their interests.

The main failing of many such designations in the past has been that they have not been sufficiently critical or objective, and elevation of common or 'less worthy' species to protection in this way has reduced the credibility of protective legislation in the eyes of many informed naturalists and biologists. Legislation or regulation must be seen to be, and accepted as, responsible and worthwhile. However, it is perhaps inevitable that some tendency to exaggerate conservation needs may occur. As Taylor (1993) commented 'we do not understand the true relative importance of various species so we are afraid to assign any a low value'.

Setting of priorities for species in conservation (Chapter 4) usually incorporates consideration of abundance (rarity) and the degree of threat, that is of status and potential harm. These have been used, through the IUCN, to designate a number of categories for the compilation of advisory lists of 'threatened taxa' of various sorts (p. 169). Such lists, including red data books, have sometimes been adopted uncritically for transfer to form the basis of laws or regulation, a purpose for which they may not have been intended initially. Many species included in such lists, for example, are unlikely to be sought or captured to any significant extent, and are included there because of the need for protection of habitat, rather than for protection of individuals from exploitation. The purpose of such lists is, therefore, usually much broader than for nominating legislative priority taxa, and they range from 'example lists' to more comprehensive regional fauna appraisals. Use of the former category for legislation without any further qualification can result in other species, in at least equivalent danger or meriting equal or more conservation need, being ignored.

Ideally, legislation for protection of priority taxa should be accompanied by provision for

Table 6.2 Threatened invertebrates listed on the 1990 IUCN *Red list of threatened animals*

Taxon	Number of species /subspecies listed	Notes
Cnidaria	*c.* 173	Uncertainty over numbers of coral species
Platyhelminthes	4	All free-living planarians
Nemertea	10	
Mollusca		
Bivalvia	109	Many mussels and clams
Gastropoda	>390	Some uncertainty over species numbers
Others	1	A European sea slug
Annelida		
Polychaeta	1	Palolo worm, Pacific
Hirudinoidea	1	Medicinal leech,
Oligochaeta	140	European earthworms
Arthropoda		
Merostomata	4	Horseshoe crabs
Arachnida	>18	>16 spiders, 1 pseudoscorpion, 1 harvestman
Crustacea	127	
Insecta	>1225	
Onychophora	(all) (*c.* 100)	Listed as Peripatidae: 'Peripatus' (10 genera) Peripatopsidae: 'Peripatus' (12 genera)
Echinodermata		
Echinoidea	2	

study to clarify their status and habitat security for them while this is accomplished. In such contexts, protective legislation can be an important facet of species management. The sheer number of species meriting priority may overwhelm logistic capability, however, and some workers query strongly the value of this approach for invertebrates.

RANKING SPECIES FOR LEGAL PROTECTION

The IUCN invertebrate red data book (Wells *et al.* 1983) was a milestone in invertebrate conservation. It provided, for the first time, a constructive focus of the sort that had long been taken for granted by conservationists working with vertebrates. It dealt with a range of species and higher taxa, and summarized many of the general issues involved. Several more local red data books (RDBs) have appeared since then—

two for Britain (insects; invertebrates other than insects) summarize priorities in the best-known local invertebrate fauna, but others appear to have been less selective in the taxa that are included. Although RDBs are advisory, indicating existing or potential losses and thereby seeking to indicate the taxa that need priority conservation attention (Fitter and Fitter 1987), they and other listings have sometimes been used as the basis for formalizing protective legislation, with little or no critical 'filtering' taking place. They thus may 'pre-empt' priorities, and the presence of RDB-listed species on a site can more-or-less automatically enhance its conservation status, and may be usable as a basis for seeking conservation funds. Likewise, inclusion of species on the IUCN *Red list of threatened animals* (IUCN 1990*a*) (Table 6.2) signals priority for conservation attention. 'Red data book species' are discussed frequently in the conservation press, and such avenues for communication are increasingly important in fostering public interest (Ferrar 1989). Inevitably, RDBs favour taxa for which information is to hand, and most are not comprehensive.

Not all species included in a RDB are of equal priority. Following the more general scheme adopted by IUCN, one role is to attempt to indicate priority by allocating species to one of several categories. These have been used more widely for other biota, but have shortcomings when applied to many invertebrates because it is possible only rarely to determine a species' abundance, distribution, and conservation status except in rather broad terms. Many invertebrate species cannot thus be ascribed unambiguously to one or other of the categories used (Appendix 1), but can be classed merely as 'inadequately known'.

It is, sometimes, possible to be more categorical. For a global appraisal of the swallowtail butterflies, Collins and Morris (1985) were able to allocate 99 (of the total 573 species in the family) reasonably firmly to one or other threatened category but nearly the same number to 'inadequately known', reflecting the need for more detailed field surveys to determine their status more precisely. The categories noted in Appendix 1 are designed to rank the degree of urgency for conservation and to serve also as a 'warning' to monitor species in each to detect whether their need increases with accelerated or increased intensity of threats. The 'extinction' category is difficult to apply for such generally inconspicuous animals, and the 50 year period since the last field sighting is more easily applicable to organisms with much longer generation times than those of most invertebrates.

Mace and Lande (1991) have proposed an alternative system for ranking species for conservation priority. They suggested only three categories, and based these on estimation of probability of survival. This approach presupposed the application of population viability analysis (PVA) to the taxa, the determination of effective population size, and the availability of sound demographic or life-table information. It will be extremely difficult to apply to most invertebrates, because of the lack of such information and the difficulty of gathering it. Even if it proves practicable for a few species, it will be unwise to extrapolate, even to closely related species or, indeed, to geographically disjunct populations of the same species. Local factors have been shown repeatedly to influence invertebrate population dynamics substantially.

Thus, even obtaining a consensus view on the conservation status of a species can be difficult, despite the appeal of apparent objectivity by categorizing them in some way according to status.

The categories allocated by Wells *et al.* (1983) represent a range of levels of certainty, and the volume illustrates well the problems involved in doing this. The coverage could be extended many-fold as the invertebrates included are usually only examples of a much broader conservation need in many phyla. However, getting agreement over what taxa should be included in any such listing is frequently difficult: many people have expressed doubt over the value of listing enormous numbers of invertebrates which may be at risk in some way, because such lists rapidly become far too long to 'handle' with limited logistic support.

For Sweden, for example, Andersson *et al.* (1987) presented a 'preliminary list' of 786 taxa which they regarded as threatened. Most of these are terrestrial, reflecting the more limited knowledge of the status of many freshwater taxa. Of these, 94 (1 mollusc, 93 insects) were extinct, and the larger categories were 'endangered' (150 species), 'vulnerable' (241), and 'rare' (261): the remaining 22 species were in need of monitoring and care. By far the predominant representation was of Coleoptera (303 species) and Lepidoptera (251 species), reflecting both their diversity and the levels of knowledge available.

In Britain, the insect red data book (Shirt 1987) included 1786 taxa, some 14.5 per cent of the total fauna of the eight orders included, with comparable figures for Coleoptera (546 species) and Lepidoptera (122); substantial portions of the fauna of other orders were also listed (Table 6.3). For non-insect invertebrates (Bratton 1991), the predominant group was the spiders (86/122 species included in the volume, or 13 per cent of the British spider fauna, of conservation concern).

Several broader-scale attempts have been

Table 6.3 Summary of insect species (and subspecies) and categories included in the British red data book for insects (Shirt 1987)

Order	Category					
	Endangered	Vulnerable	Rare	Other	Total	% Total fauna
Odonata	4	2	3	0	9	22.0
Orthoptera	3	2	1	0	6	20.0
Heteroptera	14	6	53	7	79	14.6
Trichoptera	9	4	18	2	33	16.6
Lepidoptera	27	22	55	21	122	33.1
Coleoptera	142	84	266	60	546	14.0
Hymenoptera (Aculeate)	37	12	97	18	164	28.3
Diptera	270	226	328	3	827	13.8
Total	506	358	821	111	1786	14.5

made to determine the status of invertebrates: early ones were by Heath (1981), who discussed the threatened butterflies of Europe, and the European Invertebrate Survey, with the major aim of determining the distribution of the vulnerable taxa (Heath and Leclerq 1981) (p. 169). An appraisal of threatened non-marine molluscs in Europe (Wells and Chatfield 1992), included discussion of about 104 species, from many countries. Stroot and Depiereux (1989) emphasized that most 'red lists' and their equivalents have been derived empirically and that, where the data permit, it is generally valuable to determine objectively the extent of decline by comparing older and more recent data as a method of inferring the vulnerability of each particular set of taxa. This would render listings more credible, as the data would be based on more rigorous methodology than has been employed in many more subjective early assessments. However, it is clearly impracticable for taxa for which historical records are unavailable (that is, in practice, most invertebrate groups), but the approach was exemplified by Stroot and Depiereux (1989) for the Belgian caddis flies (Trichoptera).

For a fauna of 200 species, nearly 40 per cent of records were from before 1950 and the rest were more recent. Differences in incidence were ranked, and showed that 22 species were declining. Nineteen of these were included in the 1987 red list (Stroot 1987). This approach is valuable

in the context of assessing decline, but not for taxa whose status does not change—even though they may be rare and their vulnerability increased through anthropogenic effects. Habitat, or other, 'subsets', such as separation of lentic and lotic species in this example, can be ranked separately to provide more 'focused' assessments if needed. With adequate knowledge, members of a higher taxon of invertebrates can indeed be ranked soundly. In a related example, Table 6.4 indicates the parameters used to assess European dragonflies in this way.

LEGISLATION TO CONTROL EXOTIC SPECIES

The increasing frequency of citation of exotic species as threats to invertebrates (p. 54) has emphasized the need for effective quarantine measures to attempt to control and limit their spread, as international traffic and trade continues to increase in volume and frequency. The great diversity of modes of transport and biological contexts for exotic species renders such regulation legally and practically complex. It is best illustrated for invertebrates by briefly noting some examples, and that a number of national and international steps are receiving more attention than ever before.

Table 6.4 Ranking species for inclusion in lists of threatened taxa: the parameters used for Odonata in Europe (from van Tol and Verdonk 1988)

Parameter	Likelihood of inclusion	
	Lower	Higher
Intraspecific variation	Small	Large
Species range	Large	Small
Position of Europe in species range	Edge	Centre
Species endemic to Europe	No	Yes
Population density	High	Low
Population trend: 20th century	Increase	Decline
Trophic level of biotopes frequented	Eutrophic	Oligotrophic
Habitat range	Eurytopic	Stenotopic
Resilience to environmental changes	High	Low
Dispersal power	High	Low
Potential population growth	High	Low
Ecological strategy	r-strategist	K-strategist
Conspicuousness	Small	Large
Effect of construction of artificial biotopes	High	Small

The introduction of exotic marine species to the Great Lakes of North America (p. 56) has led to considerable debate over modes of preventing ballast-water introductions, through earlier replacement of ballast water by ships. The legal background to this is complex (Bederman 1991). Australia was the first country (1989) to introduce voluntary restrictions that require a master intending to discharge ballast to certify that

(1) ballast water was taken from an area free of dinoflagellates (some of which can cause major mortality in fish and other marine animals); or

(2) that reballasting occurred in open tropical waters.

If this is not done, a ship is required to submit a health clearance certificate before ballast water can be discharged. These measures were undertaken after extensive discussion with the shipping industry, and the level of agreement and compliance has been high (Bederman 1991). Ballast exchange is now required also for ships entering the Great Lakes. A range of different ballast water treatments is possible, from use of biocides or installing filters, to mid-ocean exchange between ships, or using permanently stored supplies of chemically treated water kept in shore tank reservoirs. These methods vary in cost and efficacy.

Exotic species, once introduced, are usually very difficult to eradicate—many invertebrates are regarded, in practical terms, as impossible to remove once they have arrived. Prevention or control of entry thus becomes of paramount importance. There is no doubt that lack of quarantine regulation, or its ineffectiveness, has resulted in harmful introductions and a vast array of accidental or neglectful introductions of unknown significance to many parts of the world. Some of these appear to be undertaken naïvely and reflect unnecessary importation of organisms associated with commodities. Howarth (1986) cited several cases for Hawaii, which he believed merited more detailed sanitation, perhaps especially in relation to such vulnerable and isolated island systems: importation of dried cow dung for 'cow-chip throwing contests', polo ponies transported without quarantine, untreated Christmas trees from the American mainland, cut flowers from the Philippines, plant propagules for the floristry trade, plants and animals for the aquarium and pet markets, are all regarded as 'high risks', and Howarth stated that several pests have been introduced recently by each of these routes. Disease transfer is also a factor that can result from careless importations: the crayfish plague fungus (p. 134) was introduced to Europe with resistant American crayfish, and has devastated native crayfish wherever alien stocks are cultivated. However, the plague was probably introduced to Ireland (where there are no American crayfish) on dirty fishing gear (Reynolds 1988). A strong recommendation from the International Association of Astacology (1987) was, in part, 'that Governments find the means to stop the importation of living crayfish into their countries for any purpose . . . except for governmentally-approved research, reintroductions and introductions' and 'those Governments should be responsible for assuring that such living crayfish are parasite and disease free'.

Effective quarantine for small invertebrates is extraordinarily difficult. That for plant pests, for example, has to be based on risk assessment, as the logistic capability of most national quarantine agencies is insufficient for total inspection at each of the numerous ports of entry. Procedures such as pesticide 'disinsection' of incoming aircraft, and sanitation of airport surrounds (to remove 'introduction sites' such as mosquito breeding ponds), undoubtedly help to reduce the frequency of introductions, but are not totally effective (Russell *et al.* 1984). Aircraft have been particularly important in enabling rapid passage of many animals and plants between northern and southern temperate regions without them having to contend with the contrasting regimes of the tropics. This syndrome is very different from the more usual gradual expansion of an invasive species' range within a geographical area, though the latter becomes of major concern after an exotic species has been introduced (p. 55).

It is also necessary to plan and ensure the safety of deliberately introduced taxa before they are released. Knowledge of the species in its native habitat is a prime need, not only to clarify its basic biology but also to attempt to predict changes in the proposed new environment. Several years of study may be needed, with the results to be appraised by an expert committee. Once any decision to proceed is taken, brood stocks should be established in quarantine in the receiving country and, if no

disease problems emerge, only offspring should be released. Indiscriminate transfer of organisms between oceans, freshwater bodies, or land masses must be avoided wherever possible.

Similar caveats apply to introduction of biological control agents. Cullen and Delfosse (1985) discussed development of 'enabling legislation' for introductions as exemplified by the Australian Biological Control Act (1984). Assuring safety, in as much detail as can be achieved, is a prime concern, though some ecologists would claim that it is rarely, if ever, possible to predict fully the outcome of a deliberate introduction because any combination of alien species and receiving environment is unique (New 1994*a,b*).

UNIFORMITY

In attempting to devise and apply any kind of 'regulation' of the kinds discussed in this chapter, practical problems arise because of:

(1) the effects of scale—whether local, regional, national, or global, and

(2) different modes of assessment or ranking of priorities in different political units, such as adjacent states or countries within the range of the species concerned.

As examples:

1. Protective legislation for insects varies widely between different countries of Europe, or regions of the same country (Collins 1987*b*).

2. Many states of the USA have developed their own systems in order to set priorities in conservation action (Master 1991).

3. The minimum size for capture of crayfish can differ between different states of Australia.

The system promulgated through the United States Nature Conservancy via the Natural Heritage or Biodiversity Network (discussed

Table 6.5 Elements for global ranking of biological elements (e.g. species) (after Master 1991)

Level	Criterion
G1	Critically imperilled globally (typically five or fewer occurrences)
G2	Imperilled globally (typically 6–20 occurrences)
G3	Rare or uncommon but not imperilled (typically 21–100 occurrences)
G4	Not rare and apparently secure, but with cause for long-term term concern (usually > 100 occurrences)
G5	Demonstrably widespread, abundant, and secure
GH	Historical occurrence (possibly extinct; still searching in expectation of rediscovery)
GX	Presumed extinct throughout range
G#G#	Range ranks; insufficient information to rank more precisely
G?	Not yet ranked
G#T#	For infraspecific taxa; G applies to the full species, T to the infraspecific taxon
G#Q	Taxonomic status questionable

by Master 1991) is an iterative process giving 'ranks' to species, infraspecific taxa, or communities, based on a range of criteria, including number, quality and condition of 'occurrences' (= populations), extent of range and habitat, trends in population and habitat, threats, fragility (likely susceptiblity to threats—such as pollution), and others, in a computerized survey. Ranking on a scale of 1–5 can be made for global, national, and other scales on such criteria, and these can be combined to provide a single statement which can be updated regularly as new information is obtained. Global-level indices, to exemplify this approach, are shown in Table 6.5.

Not all 'regulations' in management of invertebrate populations are modern derivatives. Some have evolved from more traditional appreciation of the value of sustainable natural resources. In Oceania, traditional widespread lagoon and reef tenure by local clans or families was associated with the regulation of exploiting marine resources from the beach to the seaward edge of outer reefs (Johannes 1978). People sometimes depended on the sea for a high proportion of their protein, and such regulatory measures ante-date most 'Western' parallels by several centuries. Traditional conservation measures there included closing of fishing (or crabbing) areas, allowing a portion of the catch to escape, holding excess catch alive until needed, and laws on capturing small individuals (crabs, giant clams), as well as a wider range of prohibitions for

vertebrates, especially for turtles. Much of this kind of traditional management for sustainability has broken down during the past few decades because of the impact of 'Western' factors. Johannes (1978) noted three interconnected causes:

(1) introduction of cash economies;

(2) breakdown of traditional authority; and

(3) importation of new laws and practices by colonial powers.

Effectively, these introduced commercial competition, with need for more efficient and extensive exploitation to satisfy export markets, for more expensive equipment as stocks decrease, and concentration of the enterprise to relatively few entrepreneurs rather than remaining with the traditional 'owners'. Such trends apply to any resource, but marine invertebrates are indeed among the victims. Fostering the 'profit motive' among Pacific island people was a major aim of colonial powers: without it, people could not be induced to work in plantations, and would not have cash to buy imported trade goods! All too often the need for conservation has been realized only late in such sequences, once a needed resource has become scarce or difficult to obtain. Regulation, rather than biological manipulation, is then likely to be the predominant management component or, at least, a significant one, and much protective legislation arises directly from situations of perceived scarcity or threat.

ASSESSING DISTRIBUTION

Determining the distribution of invertebrates is an integral part of assessing their conservation status and their possible need for management or inclusion in protective legislation. Although the distribution of most species is known only in the most general terms, and even the nuances of habitat distribution which determine the incidence of most species are obscure, a number of attempts to determine this more comprehensively have emerged and are a major thrust to increase understanding in many temperate-region countries. For most groups, any such attempt to map fine-grain distribution at a level of species is patently impossible, but some better-known taxa in relatively small faunas have been instrumental in indicating how the 'mapping' process may be pursued, and stimulating the development of increasingly complex databases and geographic information systems incorporating information on invertebrates.

The approach was pioneered, initially for Lepidoptera, by the Biological Records Centre in Britain, and the distribution maps produced for these and some other insect groups are a remarkable demonstration of what can be achieved for the best-known groups of invertebrates in a reasonably well-documented fauna.

This project drew initially on the mass of published and unpublished 'collector intelligence' for butterflies and larger moths, and from the data on specimens in museum collections. This source itself poses problems because, in the past, a number of collectors were reluctant to publicize 'their' favoured collecting sites for fear of losing a monopoly on specimens of rare species, and continental European specimens of rare migrants were sometimes represented as British. A number of false label sources therefore existed (Allan 1943), but most such incidences from Britain have now been clarified. More recent records (see Heath *et al.* 1984) result from direct submissions by collectors and scientists, and the third phase of data accumulation for the Lepidoptera maps involved deliberate targeted surveys to 'fill in the gaps'. The main geographical grid used for recording is a 10 × 10 km square, with provision for subdivision to 1 × 1 km squares for more local needs. This scale is still sufficient to mask great ecological variations, and to pose problems for collectors seeking rare taxa, and much of the more precise data on rare species tends to be 'restricted' rather than published routinely.

Some similar projects have been undertaken in other countries of Europe and, more sporadically, in other parts of the world, but no universal standard survey grid has been adopted. In Victoria, Australia, for example, maps of butterfly (ESV 1986) and jewel beetle (Buprestidae) (Burns and Burns 1992) distributions have utilized a 10 × 10 minute latitude/longitude grid, giving areas of approximately 18 × 15 km. In some other cases, direct plotting of latitude and longitude coordinates has occurred.

For such conspicuous and appealing animal groups, regular surveys can be organized and involve the large number of devotees to contribute to knowledge of distribution and relative abundance. The annual 'Fourth of July butterfly count' organised in North America through the Xerces Society (p. 167) is the best-known example. The report for each year's survey, spanning about a month around the title date, lists the species found at each site visited by observers, together with notes on relative numbers seen, weather conditions, changes of the site since any previous survey, and the names of the recorders. New sites are being added continually to this survey, and data for some sites now span more than a decade.

Such surveys, patently, are not practicable for most invertebrate groups, although it may well be possible to augment current distributional knowledge substantially with rather little effort, if public participation can be captured, and enthusiasm fostered. In Australia recently, schoolchildren have co-operated over much of the east and south-west of the country to sample earthworms in their areas, and the scheme (Earthworms Downunder) has been co-ordinated through CSIRO (Baker, personal communication; 1992 see Anon 1992*a*,*b*). About 1450 people participated, in digging up a small patch of ground near their homes, sorting and counting earthworms,

Table 6.6 Limitations on invertebrate inventory: the British Diptera (from Disney 1986*b*)

Objective:	To map distribution of British Diptera on a 10 km square basis
Assume:	1. 1000 spp./square, with species evenly distributed over square (simplification of assumption)
	2. 1000 recorders who can identify Diptera to species level
	3. Each recorder identifies four species/hour
	4. Each recorder spends 20 hours/week identifying Diptera

No. of recording units: *c*. 2000 10 km squares in UK

Comprehensive mapping requires: 1 000 000 000 identifications

Possible with above: 4 160 000 identifications/year

Therefore, need 2404 years

This is a massive underestimate of reality: *actual* level of effort suggests minimum of 10 000 years needed

trialling an identification key, and sending in worm samples for their identifications to be checked. At least 15 new species were collected during this exercise! Comprehensive surveys, even for the best-known invertebrates such as butterflies, are limited largely to temperate regions where diversity is moderate (and where historical data may allow some functional interpretation of trends over the past half century or so), but increasing numbers of 'spot' surveys, or lists of taxa from a particular site, are gradually being compiled over parts of the tropics (for example Amazonia (K.S. Brown 1991) and Borneo (Holloway 1976)), and are of immense value in helping to interpret distribution over ecologically complex landscapes. However, detailed 'grid-mapping' of the kind exemplified above is never likely to occur for most of the tropics, and for many invertebrate groups even in temperate regions.

Many specialists in any given taxon have studied 'their' organisms extensively in the field and have considerable unpublished knowledge of the status and distributions of the species with which they work. Without good 'field guides' based on macroscopic and easily discernible characters, and without vastly improved taxonomic background, much of this knowledge cannot be augmented constructively by other people.

Disney (1986*b*) showed dramatically the limitations inherent in mapping the distributions of a moderately diverse group of invertebrates, even when most species are named and the recording basis and ethic is well established. His analysis of the logistics of documenting the British Diptera

in this way (Table 6.6) demonstrated that getting even an ecologically crude 10 × 10 km square analysis is impracticable.

Much of invertebrate conservation need must be assessed without such objective knowledge, and this is undoubtedly a weakness in political negotiation, together with the impracticality of obtaining invertebrate inventories for any particular site.

Databases such as that accumulated for the British butterflies yield far more information that is relevant to conservation management than distribution alone. Historical interpretation can define periods of range expansion or contraction, analysis of label data can indicate flight phenology and its variation between years, and correlations with various weather factors, geology and vegetation can all lead to greater understanding of the species' resource needs and the factors that determine its distribution and abundance. Anthropogenic influences can also be superimposed to detect, for example, the effects of land management, vegetation change, or increased human settlement. Such correlations can provide useful leads on the kind and urgency of remedial actions needed.

INVENTORY AND SAMPLING

Two different contexts require knowledge of the invertebrates present at a particular site, in addition to assessing the status of the species themselves, as above. These are:

(1) evaluating the site for conservation value; and

(2) possible 'ranking' of that site in relation to others in a similar habitat to select the 'best' one(s) for reservation or conservation priority.

Both were noted in Chapter 4, and both absolute and comparative information on invertebrates is thus likely to be required. It is necessary, not only to know of the presence of 'notable' (priority) taxa, but also to estimate diversity and, as far as possible, to enumerate or list the taxa present. The methods by which this is done should, ideally, be sufficiently standardized to be used in a definable manner for site comparisons. For qualitative comparisons, methods and sampling intensity/frequency need to be defined. In some instances of defining site importance, merely establishing the presence of significant taxa may be sufficient justification—these may be found initially by chance, and the problem then moves to determining whether they are indeed resident on, or restricted to, the site.

The range of sampling techniques needed for even a representative invertebrate survey is considerable, especially when aquatic and terrestrial habitats both need to be appraised. Each invertebrate group has techniques developed and modified by specialists to sample it efficiently; these are of differing quantitative reliability and must be interpreted and integrated into an effective broader sampling programme, involving several complementary methods in a sampling set. Thus, a minimum protocol for terrestrial arthropods should include Malaise, flight-intercept and pan traps, as well as Berlese funnels or other behavioural extractors, and the limitations of any component method need to be understood (Biological Survey of Canada 1994).

Any survey for invertebrates must be planned very carefully (Brooks 1993), and the initial questions asked should include:

1. Why is the information needed?

2. How are the data to be used?

3. How may the information be obtained and communicated most effectively within the constraints of time and funding that apply?

In general, about five categories of invertebrate survey relevant to conservation may be undertaken. They differ clearly in purpose and emphasis, and some are combinations of various categories. The categories (Sheppard 1991) are:

1. Inventory surveys, with the aim of finding out what is there and whether there are any significant features of species incidence, diversity, or community size or structure.

2. Site comparisons, where a series of sites is compared, or ranked, in terms of their invertebrate fauna or some taxonomic segregates of this.

3. Evaluating the effect of management practices, where the aim is to evaluate the influences of planned management on invertebrates.

4. Impact assessment, requiring predictions about the effects of specified proposed activities on invertebrates.

5. Rare species survey, where specialists assess the current status of certain rare taxa on sites, perhaps in relation to definition of some critical habitat proposed for reservation or other protection.

Ecologists are often expected to produce 'definitive surveys' of a site or habitat within a very short period. For invertebrates, this is usually utterly impracticable. Any sampling programme, in order to provide reliable information, must:

(1) allow for samples at different times of a year to cater for the short apparency of many taxa;

(2) allow for short-term vagaries of weather and other factors that may influence activity (and, hence, 'catchability') of taxa;

(3) cover the range of microhabitats and environments present, each of which is likely to harbour specialist invertebrate taxa with low vagility;

(4) be sufficiently comprehensive within each of these subdivisions to give fully representative samples on each occasion;

(5) cater for widely disparate life-forms within a taxon, in order to determine which species are residents and which are 'tourists'; and, ideally,

(6) should extend over more than 1 year, to ensure replication and augmentation in time, as well as in space.

It may be important to seek advice from experts before commencing a sampling programme, and to examine any published information. These twin leads may provide suggestions on the representative fauna present, or likely to be present, in an area, and may indicate possible 'notable taxa' which might be important in assessing conservation value. Advice on obtaining identification of the material collected should also be sought: it may be necessary to preserve or prepare specimens in a particular way to ensure their optimal use and appeal to authorities. It is not satisfactory merely to 'turn up' with large collections of sorted or unsorted invertebrates and expect people to identify them—especially if the material is not in the condition needed to facilitate this. Indeed, in practice, the groups to concentrate on in a particular survey, itself designed carefully in response to answering a particular need, may be dictated in part by the availability of taxonomic expertise to define the collected material, as well as considerations of optimal indicator groups for the habitat(s) targeted.

Funding requests for any ecological survey should include provision to ensure that this support and analysis is available: all too often, in the past, the value of surveys has been diminished substantially by lack of appreciation of this need, and failure to identify beyond major group level the taxa collected. Provision should also be made for deposition of the specimens in an institutional collection, as 'vouchers' (p. 150) cross-referenced to the survey. Voucher specimens are often needed to re-check identifications later, as the taxonomy of particular groups progresses and criteria for delimiting species may change.

Details of ecological sampling methods for invertebrates are elaborated in many ecology texts (such as Southwood 1978). Two complementary approaches are:

(1) general area-based sampling, utilizing quadrats or transects to define a quantifiable and replicable part of the habitat in which organisms can be sought and counted; and

(2) more 'passive' trapping of specimens in, or passing through, an area by various forms of trap, which can be set up and inspected/emptied at intervals.

Plotless techniques involve more general collecting to determine 'what is there' but can be compared through some quantification of 'sampling effort', such as the amount of time spent in collecting.

A combination of these two approaches is common and, because any single sampling method used alone may not give representative catches, a number of different techniques should be used in any survey: a single method used to compare sites can, though, give comparative data within the limits of its efficacy. Disney (1986b) referred to the use of standard 'sampling sets', these being combinations of sampling methods designed deliberately to complement each other and furnish a relatively complete and informative sample of selected invertebrate groups. For some terrestrial insect groups, for example, he advocated a combination of a Malaise trap, yellow and white water traps, and pitfall traps, with the bulked catches of these being used for interpretation.

The sampling design for any survey will, clearly, be dictated by the demands and priorities of the survey. The general point that emerges, though, is that any reasonably comprehensive survey of the invertebrates (or of particular invertebrate groups) in a habitat or site is likely to be more complex—and, thence, more costly—to undertake than the sponsors are likely to believe. However, the information that accrues is likely to be substantial and crucial in ecological assessment.

Complete sorting of the mass of invertebrate animals obtained in a sampling programme is expensive, hence the progressive adoption and use of ecologically informative 'indicator groups' (p. 25) as a form of more rapid assessment or 'ecological shortcut' to understanding of natural communities; this topic is discussed in Chapter 2.

But it may also be desirable to attempt to interpret groups that are not obviously well known as indicators, but which reflect community condition by their diversity—by participating in a multitude of different ecological interactions with other taxa and which contain a high proportion of ecological specialists. These should include a range of different trophic groups (Disney (1986a) suggested the value of Hymenoptera and Diptera among insects of terrestrial communities), and can be selected on such criteria as:

(1)   ease and reliability of sampling;

(2)   ease of identification, including the availability of specialist advice and handbooks;

(3)   relevance to the particular study; and

(4)   background knowledge of ecological roles and responses, level of feeding specificity, and distributions.

The expectation is that sound interpretation of the ecological community from which they are collected may thus be made by reasoned extrapolation.

'Diversity' is often assumed to reflect community maturity, with the greater number of taxa occurring in more mature communities.

However, for invertebrates, diversity (be it measured as simple species richness or as a more complex measure also incorporating relative abundance) can be high even in highly disturbed communities, and it is thus necessary to know more about the species there to assess (for example) their degree of specialization or generality. Vast numbers of individuals of common species are not atypical of disturbed environments, whereas a higher species richness tends to be more typical for undisturbed communities.

FURTHER READING

Collins, N. M. (1987). *Legislation to conserve insects in Europe*. Amateur Entomologist's Society, Pamphlet No. 13, London.

Kahn, R. P. (ed.) (1989). *Plant protection and quarantine*, 3 Vols. CRC Press, Boca Raton, Florida.

Lyster, S. (1985). *International wildlife law*. Grotius, Cambridge.

Pollard, E. and Yates, T. J. (1993). *Monitoring butterflies for conservation*. Chapman & Hall, London.

Usher, M. B. (ed.) (1986). *Wildlife conservation evaluation*. Chapman & Hall, London.

7 SOME 'CASE-HISTORIES' OF INVERTEBRATE CONSERVATION

Many of the factors and processes involved in the practical conservation of invertebrates have been discussed in earlier chapters. In practice, these need to be integrated in various ways (such as by the schemes noted in Chapter 4), and the purpose of this chapter is to demonstrate how this is generally being done in this largely 'untried' science by briefly discussing a number of examples. The examples are from a variety of taxa and ecosystems. They demonstrate the variety and depth of information and understanding needed to conserve any particular species. However, other relevant case histories focus on 'faunas' (p. 142) and a discussion of cave invertebrates is included to add further perspective to the emerging discipline.

However, two significant general points need emphasis.

1. The cases discussed have a strong bias towards the temperate regions of the world, reflecting the culture of species-focusing in conservation and the support and funding to be able to pursue this. Essentially, the availability of such cases reflects capability rather than simply need, and enormous numbers of invertebrates in other parts of the world—where human pressures are severe but fostering interest in the natural environment remains a massive challenge—are equally deserving. Wells *et al.* (1983) noted that 'documentation of the conservation needs of invertebrates has been virtually restricted to the Nearctic, Palaearctic and Australasian regions' and 'The greatest concentrations of species currently at risk probably occur in the tropical countries of Africa, South America, and South East Asia'. This scenario remains as true a decade later, with the major exception of increased documentation from South Africa, including, for example, a red data book for butterflies (Henning and Henning 1989). A number of butterflies are legally protected in South Africa, but management plans for them have not yet been effected, although current efforts are addressing the needs for some of them very constructively. There is now considerable impetus for invertebrate conservation in South Africa. However, no cases are included for the three major areas noted above. A number of invertebrates are listed on various 'protection schedules' for the regions, but species-focused invertebrate conservation has not yet gained high priority over much of South America and Asia.

2. Many highly publicized cases for invertebrate species conservation are in their early stages. Most have not proceeded far along the sequence of steps discussed in Chapter 4. Status evaluation remains the priority for most species. Management plans, recovery plans, or more general 'action plans' have been designed for species in several major groups, but most have not been completed. The best examples are for butterflies; many were summarized in New (1991a, 1993a), and only representative cases for these are recapitulated here, to emphasize their importance as flagship or umbrella taxa. For most major invertebrate groups, and for most regions of the world, there are no reasonably comprehensive or advanced cases of

species-level management programmes. Protective legislation is reasonably widespread, as part of an increasing level of conservation awareness but, for most non-commercial species of tropical invertebrates noted by Wells *et al.* (1983), and many others believed to be rare or threatened, no practical conservation measures have yet been taken. Likewise, little action appears to have been taken on species listed in a number of regional red data books and similar lists.

The cases exemplify a range of values and motivations, from maintenance of commercial resources to amenity and ethical concerns for species. Collectively, they demonstrate the commitment of different sectors of the community to this aspect of 'species importance' and also confirm the impracticability of:

(1) extending this emphasis to an adequately large selection of invertebrate species; and

(2) making it a major conservation emphasis in many less-developed countries.

For many phyla, no species have yet been addressed, but other insects and molluscs have received similar attention.

Each case is discussed briefly in relation to threats, significance, conservation measures instituted or needed, and the real or likely outcomes from these. Some are based on single synoptic review accounts, and references cited for each give greater biological background.

FRESHWATER TAXA

Spengler's pearl mussel, *Margaritifera auricularia* and the freshwater pearl mussel, *M. margaritifera*, European freshwater molluscs

These two related species are among the molluscs of greatest concern in Europe, where some 200 endemic non-marine molluscs are now of conservation concern (Wells and Chatfield 1992). Whereas *M. auricularia* has been reduced to a single population, *M. margaritifera* still occurs in many countries, where it is declining

strongly and frequently assessed as vulnerable or endangered. It is apparently secure in North America.

The large unionid *M. auricularia* was once widespread over most of western Europe and Morocco, but is now verging on extinction and is known from only a single extant population, in Spain (Altaba 1990). Decreasing abundance has resulted directly from human exploitation, predominantly for its shell and pearls, over the period from the Neolithic to the present. In common with other freshwater mussels, pollution of rivers (including accumulation of pollutants in sediments) and impoundment have also contributed to decline. Most European rivers have now undergone substantial change, and cannot support such sensitive organisms.

The present population of *M. auricularia*, in the lower Ebro, Catalona, and associated irrigation channels, is under further threat. Irrigation practices have led to lining of channels with concrete and, where mussels were not buried directly by this, they were largely dredged out and discarded with sediments. Input of sewage and toxic wastes to the river continues; much of the riparian forest that formerly provided shade and protected shores from erosion has been removed; water diversion for urban use and irrigation is planned; exotic fish have been introduced; and shells are still desired by collectors.

The life cycle of such mussels necessitates a larval (glochidial) phase attached to a fish or amphibian: although details for *M. auricularia* are not known, it is probable that the main host is a sturgeon (*Acipenser sturio*), a fish which is itself endangered (if not already extinct) in the Ebro system. It is thus possible that the mussel population is well advanced on the road to extinction by lacking any significant recruitment.

There is clear need for practical conservation measures and investigation of stocks if Spengler's pearl mussel is to persist. It is listed in the Bern Convention, and guidelines for management accepted by authorities of the Ebro Delta Natural Park are as follows (after Altaba 1990):

1. Restrict collecting: permits needed for any freshwater mussel collections, and these to be restricted to researchers and wildlife managers.

2. Research on population status, life cycle and ecology of *M. auricularia*.

3. Protect fish hosts, perhaps modifying fishing calendars to prevent removal of any fish during the seasonal pattern of carrying glochidia (this is yet to be determined); possibly reintroduce the sturgeon.

4. Protect key habitat, especially specific Ebro sites already under human pressure.

5. Inclusion in environmental education efforts: preparation of pamphlets and posters recommended.

6. Establishment of captive breeding colonies once sufficient knowledge has been accumulated to undertake this.

7. Possibility of future translocation of individuals to unoccupied sites in the same drainage area.

8. Conservation of water quality, with regulations on water discharge, preservation or reconstruction of riparian forests, and regulation of impoundment release.

The related Central European pearl mussel, *M. margaritifera*, is also threatened with extinction (Bauer 1991, Wells and Chatfield 1992), with much of the decline due to effects of eutrophication on young mussels, probably through increased sedimentation. Many populations lack young mussels at present, and the long-lived molluscs (up to around 100 years) do not reach adulthood until they are about 20 years old. A long recruitment period is, therefore, needed. Pearl-fishing has also engendered decline in northern parts of Europe (Bauer 1988).

Margaritifera margaritifera is the best-documented threatened mollusc in Europe, and the many reports and recommendations for improved management and protection have been summarized and interpreted by Wells and Chatfield (1992). They overlap with those for *M. auricularia*, but are more complex because of the greater variety of habitats needing to be incorporated. The main factors involved for practical conservation are:

1. Reduction of eutrophication and pollution is regarded as the key requirement for long-term survival.

2. Create reserves in unpolluted areas, where possible, and draw up management plans for rivers in several specified countries.

3. Prevent illegal fishing, and implement strict controls on licensed fishing. Controls drawn up for Britain could apply to the rest of Europe; if control is difficult to enforce, laws on pearl-fishing may be needed.

4. Control engineering activities that alter river banks, flow patterns, and sedimentation. Possible liming of streams to halt acidification may be needed in places.

5. Restock rivers with the predominant host, brown trout (*Salmo trutta*), and extend closed seasons for the fish, to protect them, where appropriate.

6. Re-introduce mussels as glochidia on host fish.

7. Prevent introduction of exotic salmonids which might compete with native host species.

8. Initiate and extend surveys in many countries.

9. Improve national protective legislation where necessary.

10. Co-ordinate the various research projects under way in Europe and set up a centralized database for for the mussel.

For both these mussels, the management plans suggested draw on a considerable knowledge of the species' biology and of the wide range of threats that have been defined.

The medicinal leech, *Hirudo medicinalis*, in Europe

Use of leeches for bloodletting in medicine extends from at least the fifth century BC

(Minkin 1990). Supply of leeches became a major industry in the nineteenth century and supplies were becoming depleted then in many parts of western Europe (Wells *et al*. 1983). Importations from other parts of the continent became common: France, alone, imported around 4.5 million leeches in 1833, for example (Minkin 1990).

Leeches still have important roles in medicine; for example, in the control of swelling after microsurgical reattachment of severed digits. *Hirudo* has also been found to contain important anticoagulant compounds, resulting its their demand by the pharmaceutical industry. Some such compounds are proving amenable to genetic engineering approaches for commercial development. Leeches are also being used as important research tools in studies on the functioning of the nervous system.

Intense collecting pressure on *H. medicinalis* has been exacerbated by reduction of habitat. Draining and loss of marshes in the native European range (from western and southern Europe to the Urals and eastern Mediterranean) has resulted in the decline of frogs, which are important hosts for young leeches. Reduction in the practice of watering cattle and horses (formerly important leech hosts) at ponds and 'natural' water bodies has also occurred as farming practices change.

Recognition of decline led to legislative attempts to control the leech trade from the 1820s on by several European countries (Wells *et al*. 1983), variously involving prohibition of export (from Hanover in 1823 and Sardinia in 1828), licensing areas to particular dealers (Austria in 1827), and imposing closed seasons or tariffs on exported leeches (Russia in 1848). *Hirudo medicinalis* is now on CITES Appendix 2 (p. 118) and is the subject of attempts at captive breeding (Sawyer 1976) in various parts of the world, following early (and economically unsuccessful) attempts during the nineteenth century in France, Germany, and the United States. This strategy could eventually replace field stocks for medical and commercial needs, and possibly provide leeches for release to the field if habitats can be safeguarded.

The noble crayfish, *Astacus astacus*, in Europe

This formerly abundant European crayfish, valued for food, has declined markedly since the middle of the nineteenth century and has been eliminated from much of its former range. Crayfish plague (below) has been the main cause of this, but *A. astacus* is also very sensitive to pollution, and the females accumulate DDT easily. This probably influences their long-term reproductive success. Other aspects of pollution are also detrimental. Acidification of lakes results in increased vulnerability, as 'shells' are soft for longer periods, and a decline in numbers of offspring.

Illegal or overexploitatory fishing, predation by eels, and possible competition with introduced crayfish have also been implicated as factors causing decline in particular populations (Wells *et al*. 1983). Habitat disturbance, though not as predominant as for many other aquatic taxa, occurs also throughout the species' range.

Crayfish plague is a fungal disease caused by the oomycete *Aphanomyces astaci*, and apparently developed in America, as American crayfish are largely resistant (Unestam 1973). The disease reached Europe around 1860, and spread rapidly. All European countries are affected and, largely as a result of the disease, the noble crayfish was assessed as 'endangered' in Wells *et al*. (1983).

Conservation measures taken have largely involved closed seasons, with the details differing among different countries, and minimum size levels for crayfish capture (of 9–12 cm, variously). However, more integrated conservation strategies are being developed. In Norway, for example (Taugbol *et al*. 1993), regulatory management strategies comprise:

1. Legislation—a legal catching season of only about 6 weeks, a minimum mesh size of 21 mm in traps or nets, a ban on SCUBA-captures, a minimum legal size of 95 mm total length, and banning of 'stocking'.

2. Regulation to prevent the spread of the disease—no importation of unboiled freshwater crayfish into Norway for any purpose, including the aquarium trade; no release or caging of crayfish outside the locality where they were caught; prohibition of discard of dead or diseased crayfish into waterways; disinfection of catching gear and cages between seasons, and between use in different watersheds; no used gillnets or catching gear are allowed to be imported; boats and other gear must be thoroughly dry before use in other watersheds, and water containers must not be emptied into another watershed.

In addition, attempts at re-establishment by translocation from other field populations have been made and will be monitored.

Recognition of genetic differences between populations has also led to attempts to conserve some plague-threatened populations by transferring part of the population into suitable crayfish-free lakes outside plague areas, so that these act as a living genetic resource available for re-stocking the original sites, if necessary. Five populations had been treated in this way by October 1991.

Some efforts to restrict the spread of the plague have been made by 'upstream barriers': such structures as locks may serve to retard upstream spread which would otherwise be certain to occur.

This Norwegian strategy recognizes that no importation of plague-resistant crayfish is possible, and the emphasis is on safeguarding the remaining native stocks. Aquaculture for local restocking purposes is reasonably widespread.

A 'living fossil', the damselfly, *Hemiphlebia mirabilis*, in Australia

Hemiphlebia mirabilis (the hemiphlebia damselfly) is a tiny (about 22 mm wingspan), endemic Australian damselfly, which is the sole member of the superfamily Hemiphlebioidea—a living fossil, with unique adult and larval features, and with strong affinity with some Permian taxa.

It was known until the early 1970s only in parts of the Goulburn and Yarra River valleys in Victoria but was believed to have become extinct there later that decade. The requisite habitat was one of seasonally flooded reedy lagoons in those areas, but most flood plains had been destroyed by agriculture and intrusion of cattle into the small remnant waterbodies.

However, a large colony of *Hemiphlebia* was discovered by Davis (1985) in the Wilsons Promontory National Park in southern Victoria, and its biology and status there was appraised by Sant and New (1988). Fears for the safety of this apparent relict population occurred when the swamp harbouring the largest portion of the colony was burned in 1987 (as an accident from control-burning of nearby heathland to control succession), which increased access to the swamp by cattle and resulted in substantial trampling when water became scarce in late summer, as well as causing direct damage to the habitat.

At that time, the mown roadside firebreak strip remained relatively lusher and fire-resistant: mowing over part of the swamp (which would normally be regarded as destructive) in this case might have created inadvertently a reservoir habitat for *Hemiphlebia*. Numbers of adult damselflies were very low over the next two flight seasons, but increased thereafter. With the co-operation of the National Parks authorities the swamp was fenced after the burn to exclude cattle in subsequent seasons, and the mown area (about a quarter of the total) was left undisturbed to facilitate regeneration. Numbers of damselflies have continued to increase at this site, and the 1987 burn area has been recolonized.

In 1992, *Hemiphlebia* was observed in small numbers in the Goulburn Valley (Trueman *et al*. 1992) and also in one site in northeastern Tasmania. In the following season it was found on Flinders Island, in the Bass Strait between Wilson's Promontory and eastern Tasmania (Endersby 1993), so that several geographically discrete populations are now known. Its future now seems more secure, but the fortunate discovery of the species within a national park, and its undoubted susceptibility there for several years, rendered the case unusual.

The IUCN's Odonata Specialist Group had

already placed *Hemiphlebia* as its highest priority in world odonate conservation but, although it may still be vulnerable, recent discoveries have effectively removed it from the 'endangered' category to which it was allocated formerly. *Hemiphlebia mirabilis* is an example of a species for which searches were stimulated by its designation as a protected species under Victoria's Flora and Fauna Guarantee Act, and whose biology has consequently become better understood. Further survey for colonies in the complex billabong/swamp system of the Goulburn Valley is now a priority, together with assessing the likelihood and suitability of the species for translocation if this should prove necessary. Increased security of the Wilson's Promontory sites has come from the phasing out of cattle-grazing leases in the park.

TERRESTRIAL TAXA

The edible or Roman snail, *Helix pomatia*

Helix pomatia, the largest European snail, has declined substantially over much of its range and is regarded as 'rare'. The prime cause of this seems to be overexploitation for food—in sparse populations by individual people rather than for major commercial enterprises (Wells *et al*. 1983). It has secondary use in education as a large 'typical' snail amenable to dissection and experimentation. In Britain, it takes it epithet of 'Roman' because it is believed that it was introduced first as food by the Romans; there have been many subsequent introductions (Pollard 1975).

Helix pomatia is widespread and is collected for export in many European countries. Wells *et al*. (1983) cited, for example, Switzerland importing 236 609 kg of *H. pomatia* in 1980 (90 per cent of this from Hungary). Poland exported 40 million snails between 1951 and 1961; Hungary exported 40 million in 1974 alone (Welch and Pollard 1975) and Germany, 48 million from 1971 to 1973.

Declining populations tend to have a high proportion of (old) large snails, which are vulnerable to collectors, and collecting of such specimens preferentially for scientific research has also led to depletion of some populations. Change of agricultural practices has progressively alienated some habitats, such as by encroachment on calcareous grasslands in Britain (Pollard 1975), and Wells *et al*. (1983) also cited pesticides, fires, road-building, housing developments, and loss of woodlands as having had serious effects on the snail in various places.

Conservation programmes for *H. pomatia* must, therefore, seek to control collecting and conserve habitat, and perhaps to utilize captive breeding to satisfy commercial demand for food snails. *Helix pomatia* occurs on a number of reserves, but the degree of effective protection is sometimes uncertain. Controls on collecting can include closed seasons, usually in spring and early summer, the main breeding season. In some places, a short *open* season is allowed, and many areas (such as 18/26 Swiss cantons) prohibit any collecting by declaring the species 'protected'. Individual size limits (usually prohibiting the take of snails of shell diameter less than, variously, 28–32 mm) and daily bag-limits (Italy: Trento permits a daily catch of not more than 1 kg/person, with no nocturnal collecting allowed) occur. In many parts of Germany, collection site rotation is practised, whereby a given site can be exploited in only 1 of 3 years.

Recommendations have been made, variously, for increasing the length of such cycles to 4 or 5 years, increasing minimum size levels, more stringent quota controls on collecting, and implementation of further closed seasons. Because of the snail's poor natural dispersal powers, translocation of snails to new or rehabilitated sites could be practicable, and more detailed ecological knowledge is needed.

Captive breeding is considered to be commercially unrewarding because of the slow growth rates of the snail (a minimum of 2 years, and perhaps up to 5 years, to maturity, although the eggs hatch after 3–5 weeks: Pollard 1975) and low reproductive capacity, as well as incidence of disease. It may, though, be useful in rearing to 'young snails' for release, and this could be a useful management tool.

Recent trends to farm *Helix aspersa* as a substitute for *H. pomatia* as a gourmet food might help to reduce the collecting pressure on the latter species.

However, Welch and Pollard (1975) suggested, whimsically if hopefully, that the best solution for *H. pomatia* may be one reported in a Swiss newspaper: restaurants serving mock snails made from lung tissue, suitably disguised in a delicious sauce!

The giant Gippsland earthworm, *Megascolides australis*, of Australia

The giant Gippsland earthworm, one of the largest terrestrial invertebrates, occurs in Victoria, Australia, and specimens up to 3.6 m in length have been reported in the past. The average size of present-day individuals is much smaller, but still an impressive 1 m long (Yen *et al*. 1990; van Praagh 1992). The worm's range has apparently declined considerably during the period of intensive European settlement and agricultural conversion of Gippsland. *Megascolides* now has a very patchy distribution over around 100 000 ha of the Bass River Valley. Most of the worms occur close to watercourses or soaks.

Megascolides, despite its size and longevity, is a difficult invertebrate to investigate. Its subterranean habit makes estimation of population size extremely difficult, and the worm is hard to study quantitatively. Sampling difficulty is enhanced by the worm's fragility: it breaks easily. It also develops slowly, and attempts at laboratory rearing have not succeeded fully. Possible declines (and much of the historical data on its abundance is somewhat anecdotal: Smith and Peterson 1982) are associated with ploughing, use of superphosphate or other agricultural chemicals, and replacement of dairying by crops. The cocoons may be subject to desiccation in cultivated soils.

Megascolides australis was listed as 'vulnerable' by Wells *et al*. (1983). It is accepted as an important 'emblem' locally, and an annual festival (the 'Karmai Festival' at Korumburra) is named for the worm. Most colonies known are on public land, but one population is included in the Mt.

Worth State Park (Smith and Peterson 1982; van Praagh *et al*. 1989).

Megascolides australis exemplifies well the difficulty of obtaining fundamental knowledge on the biology and conservation status even of large 'charismatic' invertebrates in some environments, even with a high level of public sympathy and support. Studies on the worm have been supported for several years, and knowledge has indeed increased substantially (van Praagh 1992). However, a detailed conservation strategy is still to be formed and the most important interim measure is to avoid further disturbance of known centres of abundance, for example by changes in drainage patterns that could increase soil acidity, further vegetation clearing, and use of agricultural chemicals.

The Uncompahgre fritillary, *Boloria acrocnema*, a butterfly on the verge of extinction?

The fritillary attracted considerable interest when it was discovered in 1978 because of its extremely narrow range on Mt. Uncompahgre, Colorado, and as the first new butterfly species collected on the United States mainland for nearly 20 years. A glacial relict species, it was known initially from a single site of about a hectare, and a second colony found in 1982 at Red Cloud Peak was also small (Gall 1984). The species rapidly attracted the attention of collectors, who were perceived as a threat as soon as 1980, and who took up to 20 per cent of the type population in 1981.

The butterfly seemed remarkably sedentary, with a dispersal range of only 50–60 m, and larvae feed only on a prostrate dwarf willow *Salix nivalis*. Legal protection was needed to control collecting, and sheep grazing (although not allowed at the type locality) could also be damaging. The type site was designated (December 1980) as a Natural Area to the Colorado State National Area Council, and *Boloria* has now been listed as 'endangered' by the US Fish and Wildlife Service. Following Gall's studies, intensive investigations of *B. acrocnema*'s distribution, abundance, and genetic variability were undertaken in 1987–88 (Britten *et al*. 1994). By

1987 it had apparently disappeared from the type locality, but small numbers had indeed persisted there and have been found up to 1993 (Britten *et al*. 1994). The butterfly has a metapopulation structure (p. 78).

Whether the butterfly can be saved, even with a heroic conservation effort, is open to question. Britten and his colleagues suggested the following management steps, with the provisos that such restricted alpine populations may, in any case, be threatened by global warming and render massive expense inappropriate, and that translocations might be possible if any strong new colonies could be found:

1. Protection of the colonies from human disturbance, particularly (Red Cloud Peak) collecting, mineral exploration, and livestock grazing.

2. Careful, non-intrusive, monitoring of population and habitat trends.

3. Protection of all sites where the species has ever been located, and visiting these several times each flight season, in case survivors persist.

4. Diligent searches for new colonies

Queen Alexandra's birdwing, *Ornithoptera alexandrae*, the world's largest butterfly

The conservation needs of this magnificent papilionid, females of which are about 25 cm in wingspan, have long been of concern (Collins and Morris 1985, New and Collins 1991). *Ornithoptera alexandrae* is regarded as 'endangered' and is rarely seen. It commands very high prices from collectors and is known only from a small area of primary and advanced secondary lowland rainforest near Popondetta, in Papua New Guinea's Northern (Oro) Province. Its great rarity has hampered accumulation of sound biological information, but the main danger has been habitat loss through expansion of the oil-palm industry and timber logging (Parsons 1984). Until recently, it had been recorded only from 10 of the 10 × 10 km squares of the Papua New Guinea butterfly mapping grid. It is

seemingly absent from much apparently suitable habitat.

It is among the rare birdwings legally protected by a Fauna Protection Ordinance, and is on Appendix 1 of CITES (p. 117). Proposals for reserved areas of habitat (M.J. Parsons 1983, 1984) would offer substantial protection for the butterfly if they were (1) gazetted, (2) exempted from road access and logging, and (3) planted with stocks of the larval foodplants, *Aristolochia diehlsiana (= schlechteri)*. The existing Wildlife Management Area of some 11000 ha near Popondetta contains areas known to be inhabited by *O. alexandrae*. Once habitat has been secured, investigation of the species' potential for ranching is needed.

Ornithoptera alexandrae continues to be a major 'flagship species' (p. 25) for conservation in Papua New Guinea, and external funding directed toward survey and reserve establishment continues. Indeed, surveys during 1992 were especially encouraging (Mercer, personal communication) and proposals that an Integrated Conservation Development Project over about 200 km² should be excluded from logging and oil-palm establishment may hold the key to the butterfly's future, together with its incorporation into butterfly-ranching activities. Enrichment of secondary habitats by planting of larval foodplants, and, perhaps, trees attractive to adults, may also be important; if the butterfly can be sustained effectively there, rates of intrusion into primary forest habitat might be reduced. Recent (1994) initiatives, funded jointly by the Australian and Papua New Guinea governments, seem likely to foster this 'conservation through development' approach constructively and provide many people in the Oro Province with sustainable alternatives to continued destruction of forested areas.

The large blue butterfly, *Maculinea arion*, a successful international translocation?

The large blue is one of five European species of *Maculinea*, all of which are endangered by their dependence on traditional cultural landscapes which are threatened by changes in agricultural land use. Each of these species has experienced one or more national-level extinctions (Elmes and Thomas 1992), and has been the focus of intensive conservation effort.

Maculinea arion was formerly widespread in southern Britain, but declined during the nineteenth century. Concern for its well-being was strong by the 1880s, and calls for restraint on collecting specimens started around then.

As with many other Lycaenidae (New 1993*a*), larvae of *M. arion* have an intricate relationship with ants, in this case with *Myrmica sabuleti*. The caterpillars obligately spend much of their developmental period in an ant nest, after feeding initially on the flowerheads of *Thymus*: the life history is described in detail in several synoptic works, such as Heath *et al.* (1984).

Decline of the species in Britain continued during the twentieth century and nearly half of the British colonies disappeared from sites that still supported large host ant populations and many larval foodplants. These included six reserves that had been established especially to conserve *M. arion*. Between 1950 and 1972, the populations in Britain fell from around 100 000 adults to fewer than 250 individuals. *Maculinea arion* became extinct in Britain in 1979, at about the same time that a complex research programme on its ecology elucidated the factors needed to assure its survival. The history of progressive isolation of colonies engendered by change in land use was exacerbated by more subtle and intricate ecological effects.

The host ant was very sensitive to changes in microclimate and was replaced rapidly by the very similar species *M. scabrinodis* if turf height was increased by only 2 cm, resulting in a slightly cooler environment (Erhardt and Thomas 1991), so that a decline in sheep grazing, together with the reduction of rabbit grazing, resulting from myxomatosis, led to substantial loss of the key resource of this species.

Thomas (1989, 1991) showed that larval survival of *M. arion* was indeed very low in nests of *Myrmica* spp. other than *M. sabuleti* and the importance of this specificity had thereto not been suspected—indeed, this manifests in the other European *Maculinea* as well (Fig. 7.1). However, there is considerable variability in suitability of *M. sabuleti* nests to host *M.*

arion caterpillars. Many nests are too small to support even one caterpillar, and there are differences in probability of caterpillars being killed by the ants.

A suitable site for *M. arion* must be sufficiently large (several hectares) to support large numbers of ants and foodplants. Elmes and Thomas (1992) showed, further, that foodplants must be within the normal foraging distance (only about 2 m) of the ant nests. Foodplants must also be dispersed widely on the site, so that the *Maculinea* caterpillars are spread over a large number of ant nests and the risk of mortality thereby dispersed.

Since 1981, a programme to reintroduce *M. arion* to Britain has been undertaken, with prior

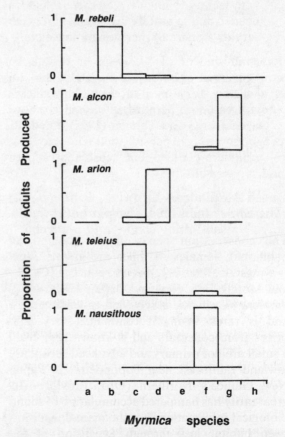

Fig. 7.1 The host-ant specificity for five species of large blue butterflies (*Maculinea* spp.), based on the proportion emerging from nests of different *Myrmica* species (a–h, not named here) in several European countries. (After Elmes and Thomas 1992.)

site preparation and protection. Details of the management programme (Thomas 1991; Elmes and Thomas 1992) included three major phases:

1. Management on former colony sites was modified, by burning and clearing scrub and increasing cattle-grazing pressure, to disturb them and facilitate the increase and spread of *Thymus* and *M. sabuleti*.

2. Selection of foundation stock, which proved difficult because of differences in biology and phenology between most European colonies (many of them small and not large enough to act as 'donors'). Only one region of Sweden was found to harbour sufficiently abundant and compatible stock.

3. Monitoring the release for several years, with facility to augment by extra releases if needed and to modify management if necessary as a result of increasing knowledge.

Re-establishment of *M. arion* in Britain has now occurred (Thomas 1989, 1991), and this is a classic story of a successful introduction attempt based on detailed autecological knowledge. As Elmes and Thomas (1992) noted, the programme was expensive, and could not have been undertaken without substantial support and sponsorship.

The El Segundo blue, *Euphilotes bernardino allyni*, a sand-dune community umbrella.

This butterfly was listed as 'endangered' under the US Endangered Species Act in 1976, at a time when the principal colony on an area of sand dunes at El Segundo, California, owned by Los Angeles International Airport (LAX), was likely to be destroyed through development of the area as a golf course. The sand dunes had been largely lost to urban development, and the butterfly occurs only in three small areas (Mattoni 1992, 1993). The largest of these (LAX) was approximately 120 hectares, and 80 hectares of this have now been put aside as a permanent reserve and the golf course (on the remaining 40 hectares) includes all its 'rough areas' as replanted butterfly habitat. Following restoration efforts, the butterfly population on

LAX rose from around 500 in 1984 to some 4000 individuals by 1990.

However, complex political negotiations were needed to achieve this. The area had been decreased rapidly since a major radar installation on a 25 hectare site purchased in 1950, and unchecked home constructions until the 1960s. Many homes were vacated from about 1965–75 because of increasingly difficult living conditions resulting from the airport expansion and building of a new runway. At this time, the proposals for recreational use began, and a series of public hearings, involving developers and those who wished the dunes to be preserved, ensued from 1981. The sand dunes support a wide array of notable plants and animals, including nine other endemic invertebrates and 15 other invertebrates restricted to southern California coastal dunes. Protection of the habitat for the El Segundo blue has therefore helped to protect many other notable taxa (Mattoni 1992) and emphasis is on restoration of the habitat to increase the pristine areas to well above the 1990 level of less than 2 hectares. The butterfly and its foodplant (buckwheat, *Eriogonum fasciculatum*) are largely restricted to land not greatly disturbed by human activity. It is necessary to protect areas from trampling, and to eliminate exotic plants and carnivores to help restore the more natural community.

Giant wetas, the 'insect mice' of New Zealand

The wetas (Orthoptera: Stenopelmatidae) are a group of large cricket-like insects endemic to New Zealand. Several of the most spectacular species, the giant wetas, are now restricted to small offshore islands (Fig. 7.2), where they are considered to be highly vulnerable. Some are known only by single small populations, and the status of the various species was summarized by Meads (1988). Seven species are listed as protected invertebrates under the NZ Wildlife Amendment Act (1980), together with an additional species which was discovered only in 1988 (Meads 1990).

Wetas evolved in the absence of ground-dwelling mammal predators (the same environment

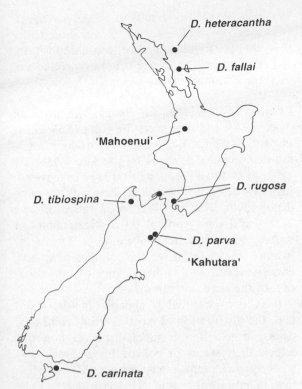

D. heteracantha

D. fallai

'Mahoenui'

D. rugosa

D. tibiospina

D. parva

'Kahutara'

D. carinata

Fig. 7.2 Distribution of the giant wetas (Orthoptera, *Deinacrida* spp.) in New Zealand. The scree weta, *D. connectens* is widespread in the South Island, and omitted from this map. The eight species shown (two of them undescribed) all have very limited distributions, some on single small islands. (After Meads 1990.)

fostered the development of flightless birds in New Zealand). Introduction of rats, including *Rattus exulans* by Polynesians and later *R. norvegicus* and *R. rattus* by Europeans, and house mice, resulted in the decline of many large New Zealand invertebrates (Ramsay *et al.* 1988), so that rodent-free islands are the last reservoirs of some taxa.

Cats have been reported feeding on the Herekopare weta, *Deinacrida carinata*, and brush-tailed possums (introduced from Australia) also feed on wetas. Exotic vertebrate predators are undoubtedly the main threat to wetas, although habitat change has contributed also to decline of some species.

The establishment of predator-free reserves is regarded as vital to conserve mainland weta populations (of the Kaskoura weta, *D. parva*,

and Wetapunga, *D. heteracantha*), and establishment of captive colonies of the various species was also high priority (Ramsay *et al.* 1988). Preventing rats from getting to several islands is vital, and Stephens Island (*D. rugosa*, the Stephens Island weta, occurs only there, on Mana Island, and, possibly, on Middle Trio Island) and Middle Trio Island are flora and fauna reserves. Wild cats have been eliminated from some other islands, and there have been moves to introduce colonies of wetas to rat-free islands in the hope of establishing new populations (Meads 1988). A third population of *D. rugosa* was established thus on Maud Island (Meads and Moller 1978).

There are some potential problems in co-ordinating optimal conservation moves for various New Zealand biota. Howarth and Ramsay (1991) noted that an endangered bird, the saddleback, had been introduced to Little Barrier Island, where the only surviving population of the protected *D. heteracantha* occurs. The bird eats wetas, among other insects.

The wetas are a flagship group for invertebrate conservation in New Zealand, and recovery plans for several species have been prepared (Meads 1987*a*, *b*, Ramsay *et al.* 1988), together with a descriptive booklet on the insects (Meads 1990). Techniques for rearing wetas in captivity have also been defined (Barrett 1991). The interim recovery plan for the undescribed Mahoenui weta incorporated a need to determine how and where new populations could be established to help buffer the species against extinction. This was advanced by studying the only known substantial population in the 240 hectare reserve purchased by the Department of Conservation in August 1990. Within that area, the weta is distributed patchily in remnant patches of gorse which may protect it against predators. A study of the species (Sherley and Hayes 1993) showed highest numbers (52 per cent of 276 individuals) in the intermediate parts of gorse bushes, with most found sitting on dead foliage and many within the reach of browsing goats. They preferred the presumably warm north- to east-facing slopes and avoided slopes facing other directions. Goat activity seems to be important in habitat maintenance and Sherley and Hayes advocated keeping the feral goat population there at its current level: a rare advocacy for what is viewed traditionally as a very harmful animal.

Cave invertebrates

Cave faunas exemplify a geographically restricted 'island' milieu of specialized ecological communities which have high levels of local endemism and which depend on a suite of specialized environmental parameters, ensuring that they could not thrive elsewhere.

Many invertebrate species are known only from single caves or cave systems in various parts of the world, and many are obligate cave-dwellers. Particular species of Onychophora, for example, occur only in particular caves in Jamaica, South Africa, and New Zealand, respectively.

Many cave invertebrates are morphologically specialized: some arthropods, for example, are pale, long-limbed and blind, unlike their close relatives which live outside caves. The problems of conserving these evolutionarily-significant faunas are indicated by the range of threats to cave ecosystems. Howarth (1981) listed the following:

(1) mining of cave materials, such as limestone and basalt, and materials (such as guano) found in caves;

(2) land use changes near caves, including urbanization, deforestation, road construction, and water impoundments;

(3) alteration of ground water flow, including flooding of caves or removal of ground water which depletes their normal hydrological regimes;

(4) pollution from waste water, road runoff, and sewage, and use of cave entrances for dumping garbage;

(5) introduction of exotic species of facultative cave animals;

(6) local extirpation of cave animals and plants, including food-source species: this is related to land-use changes noted above; and

(7) human disturbance in diverse ways: collecting specimens, trampling, and physical disturbance (especially in access passages), tobacco smoke acting as an insecticide, changes to microclimate, waste deposition, and others, which are often by-products of recreational activity.

Elliott (1991) added, for Texas, the invasion of caves by red imported fire ants (*Solenopsis invicta*: p. 62), which began to produce infestations in cave entrances by the late 1980s. By 1991, 24 of the 64 'endangered species' caves in just two Texas counties had fire ants foraging inside them. The ants feed on a range of cave fauna, and their control by pesticide baits or boiling water has been attempted. The latter is more labour-intensive for colonies near cave entrances, but avoids any pesticide contamination of the cave neighbourhood.

It is not clear what specific factors influence the distribution of most cave invertebrates, what causes a significant disturbance, how this affects their status or constitutes a threat, and what management measures would be effective in countering this (Howarth 1981). The *only* viable strategy in such instances is to protect the habitat with its community, together with nearby areas, from undue human intrusions.

Cave invertebrates have been recognized for special protection in Tasmania, and cave 'preserves' occur in many parts of the world. Howarth (1986, Hawaii) and Elliott (1991, Texas), *inter alios*, have emphasized that new cavernicolous species are being discovered continually, as their habitats become degraded or disappear.

The significance of Tasmanian cave invertebrates was appraised by Eberhard *et al*. (1991), who identified a number of sites which supported taxa that were yet to be described, or were particularly diverse. In just one system, the Mole Creek Karst region Caves, for example, many local endemic taxa were revealed with only little systematic study. Twelve caves in that system are type localities, collectively for seven invertebrate species, and one cave (Kuba Khan) had the highest diversity of invertebrates known for any Tasmanian cave: 71 species.

This system itself merited detailed biological

monitoring and development of a faunal conservation strategy. At least 15 other Tasmanian cave systems are of equivalent importance for invertebrates (Eberhard *et al*, 1991). Cave invertebrates are included in listings of protected fauna in Tasmania.

As with other such specialized habitats, informed management may well involve restricting human access to areas they would wish to visit, and which may be important in the tourism industry. There is potential conflict in gaining sympathy for conservation but also abundant opportunity to publicize and promote its worth.

MARINE TAXA

The following examples of marine invertebrates are broadly based. This reflects the general lack of knowledge of conservation status of non-commercial marine species (the few noted by Wells *et al*. 1983, are categorized as 'insufficiently known') and the lack of attention given to them. The examples also reflect possible commercial pressures, among others, as important threats to whole groups of organisms.

Sponges, a marine commercial crop

Conservation needs of marine sponges have been engendered predominantly by the desirability of about a dozen species which are exploited commercially, though other threats to Porifera do occur. Many are very sensitive to pollution (p. 50) and increased sedimentation, so that sponges may disappear from coastal areas near sewage outflows or areas of increased turbidity.

Commercial sponge fisheries in the Caribbean and Gulf of Mexico were drastically affected by a fungal disease in the 1930s and 1940s (Wells *et al*. 1983), and bottom trawling may remove sponges as bycatch (p. 69) as a component of more broadly affected benthic faunas. Some shallow-water coastal sponges may be affected by adjacent habitat modification, such as removal of mangrove swamps.

Overfishing, particularly in the Mediterranean and Caribbean areas, has been reported at intervals and these are the regions subjected to highest levels of exploitation, for export. In general, Mediterranean sponges are regarded as the 'highest quality', and good natural sponges remain highly desirable in competition with artificial products. The advent of SCUBA fishing techniques has increased collecting intensity.

Conservation of marine sponges, therefore, seeks to assure the sustainability of a commercial resource, and the steps are mainly involved with

(1) regulation of harvesting; and

(2) sponge culture increase as a basis for replacing exploitation of natural sponge populations.

Measures imposed at various times during the twentieth century have included minimum size limits, to ensure that sponges sufficiently large to reproduce were allowed to persist. The USA fisheries have a recommendation of 15 cm minimum sponge diameter, based on the belief that gamete production did not start until sponges were around 14 cm in diameter. However, in some places maturation may occur at much smaller sizes. Improvement and regulation of fishing techniques for conservation can involve closed seasons or rotational use of sites, and restriction of some indiscriminate techniques, such as dredging.

In principle, sponges are easy to cultivate, because small, cut pieces will regenerate to whole organisms when they are attached to artificial substrates and replaced in the sea. Sheltered sites are needed to avoid damage from bad weather, and harvesting can then be managed effectively in response to fluctuations in commercial need. Although up to 4–7 years may be needed before a 'bed' becomes productive, this approach could assure sustainability with minimal intrusions into wild populations. The market for sponges seems likely to continue to exceed current supplies (Wells *et al*. 1983) and also to expand geographically—particularly from the Philippines and Japan.

The largest shelled molluscs: giant clams of the Indo-Pacific

The giant clams (Tridacnidae) comprise eight species, all found in limited shallow reef areas of the Pacific, and are the largest bivalve molluscs. The larger taxa, in particular, have long been exploited by people for flesh and shells, and are susceptible to overexploitation. Local extinctions have occurred, as many reefs have been stripped of clams so that little or no natural breeding stock remains there. Recent extinctions have been recorded in several island groups. Shells are needed for sale to the burgeoning influx of tourists and are popular as ornaments, and for use in floor-tiling (Indonesia). Poaching of clams seems not to be uncommon, with several prosecutions recorded. Clams are also desired by the aquarium trade, mainly in Europe and the USA, with high prices paid for large, healthy 'exhibition' specimens.

The ecological importance of giant clams in reef ecosystems is not understood fully, but may be substantial.

Giant clams are nominally protected in many areas, such as the Great Barrier Reef, Australia, but poaching is very hard to control completely because of the difficulty of surveillance over large areas. Studies are now under way in several countries on mariculture of giant clams, and prospects for commercial-scale captive breeding seem very promising. It may, indeed, be feasible to rear clams together with the valuable mother-of-pearl yielding gastropod *Trochus niloticus* in mixed culture (see notes in Wells *et al.* 1983). This approach, together with more effective protection and control of natural populations, indicates a primary management strategy for these and other 'exploitable' marine molluscs.

Recent developments in tridacnid mariculture (Fitt 1990) emphasize the need for provision of symbiont zooxanthellae by introducing cultured or isolated zooxanthellae to clam veligers and thereby providing for the satisfactory metabolism of the clams. This also increased survival rates markedly in reared stocks.

There are three main phases in giant clam culture:

(1) a 'hatching phase' of spawning, hatching, metamorphosis, and rearing of juveniles for the first few months;

(2) an 'ocean nursery phase' of raising clams in protected trays until they are about 2 years old, a period during which their size renders them susceptible to predation by fish and crabs, and,

(3) a 'growing out phase' on the reef bottom. During this phase, increase in size is rapid, to an optimal harvesting time (for *T. derasa*) of around 5 years.

Corals: the foundation of marine communities

Reef-forming corals, in their entirety, constitute the very foundation of immensely complex marine coastal communities of incalculable fundamental value as among the most productive marine ecosystems. They have additional roles in protecting many low-lying shores from erosion, supporting fisheries, fostering tourism, and as sources of items for the 'souvenir' and aquarium trades and of building material. Some non-reef-building species are valuable for jewellery, 'precious coral' (*Corallium*) has long been used for this in the Mediterranean area, and such taxa are very vulnerable to overexploitation. This is so also for black corals (*Antipathes*) and their relatives in many tropical regions.

Control of harvesting of these is vital. Collecting is regulated, at least nominally, in many places at present, and the corals of some areas are totally protected. Minimum size limits for black corals are in force in Hawaii.

Depletion of corals seems to be particularly severe in the Caribbean area. *Corallium* spp. have been the subject of concern over scarcity in the Mediterranean, where the effects of exploitation have been augmented by susceptibility to pollution and siltation. Modification of fishing methods, including avoiding indiscriminate collecting with dredges, is needed, and the use of 'submersibles' for deep-water selective harvesting has been promoted by commercial ventures in Hawaii and Taiwan.

Table 7.1 Caribbean reef corals: some patterns of disturbance at different spatial and temporal scales (Jackson 1991)

Process	Spatial extent	Duration	Frequency
Predation	1–10 cm	Minutes–days	Weeks–months
Damselfish gardening	1 m	Days–weeks	Months–years
Coral collapse	1 m	Days–weeks	Months–years
Bleaching (individual)	1 m	Days–weeks	Months–years
Storms	1–100 km	Days	Weeks–years
Hurricanes	10–1000 km	Days	Months–decades
Mass bleaching	10–1000 km	Weeks–months	Years–decades
Epidemic disease	10–1000 km	Years	Decades–centuries centuries
Sea-level or temperature change	Global	Years	10 000–100 000 years

However, reef-building corals are subjected to a rather different range of pressures from these species (Table 7.1), associated with their ready accessibility. Pollution effects, other than sporadic extreme cases, are generally much less severe than prolonged silting, but heated effluents affect some species. Exploitation for building materials has caused loss of large tracts of reef in South-East Asia, particularly, and reefs near human settlements and tourist resorts are very vulnerable to local direct destruction through trampling (Liddle 1991), boats (dragging anchors), destruction by fishermen and souvenir hunters. This situation is unusual in that there seems to be little (if any) danger of any species of reef-building coral becoming extinct, or even threatened, because of their generally wide distribution, but they are local 'umbrella species' (p. 24) for a mass of other taxa, including some that are more localized and likely to be more susceptible. Climatic change (p. 159) may be of major significance.

Reef conservation entails control of exploitation, perhaps through regulating the access of tourists and others, and through monitoring or prohibiting trade. It is feasible to transplant stony corals, to aid in recovery of damaged reefs, and to construct 'artificial reefs' (p. 97) (Alcala *et al*. 1982). Importantly, and more difficult to achieve, control of endangering processes is needed. However, Salm (1984) emphasized that very little is really known about how to manage reefs. Poorly planned coastal developments lead directly to reef destruction by sedimentation, dredging activities, thermal pollution (from power-plant effluents), and sewage discharge. Reefs in the Red Sea have been killed by chronic low-level pollution from oil and phosphates. Even goats, through causing erosion from coastal dunes, threatened Mozambique reefs by increasing the amount of sediment on them. The diversity of threats, and their scales of influence, to Caribbean reef corals are noted in Table 7.1.

Guidelines for the design of coral-reef reserves (after Salm 1984) include:

1. The protected area should contain many different coral habitats, so that a steady and varied supply of larvae can be maintained.

2. The protected area must take into account that the reef community is not defined by the reef-edge, but extends over neighbouring habitats with which it interacts.

3. Reef management must extend beyond the protected area boundaries for reefs bordering large island or mainland coasts.

4. The core area of a coral reef protected and designed to preserve biotic diversity must be larger than a critical minimum size if all species are expected to survive.

The principle of the need to buffer protected areas adequately (Chapter 4) is universal.

FURTHER READING

Collins, N. M. and Morris, M. G. (1985). *Threatened swallowtail butterflies of the world*. IUCN, Gland and Cambridge.

Wells, S. M. and Chatfield, J. E. (1992). *Threatened non-marine molluscs of Europe*. Council of Europe, Strasbourg.

Wells, S. M., Pyle, R. M., and Collins, N. M. (1983). *The IUCN invertebrate red data book*. IUCN, Gland.

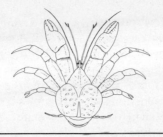

8 THE FUTURE OF INVERTEBRATE CONSERVATION

This chapter pursues two main themes: first, a summary of the major impediments to effective conservation of invertebrates and to using invertebrates in broader conservation management; and, secondly, suggestions for practical ways to advance the understanding and practice of conserving invertebrates.

DIVERSITY AND LOGISTICS

Invertebrate diversity will continue to defy attempts to document it thoroughly, but, as noted in Chapter 1, *patterns* of species richness and distribution are the template against which we can assess the need for conservation action. In essence, this equates to the need to understand some aspects of *invertebrate* biodiversity. May (1990), among others, has stressed the need to understand how biological systems 'work': how ecosystems function', and this need underlies most of the content of invertebrate conservation, with management for species, communities, and ecosystems needing sound (and often intricate) ecological knowledge for effective prosecution. Yet, this knowledge is fragmentary and even some of the rather imprecise statements on levels of species 'diversity' noted earlier are open to severe revision as knowledge accumulates. Returning to a theme of Chapter 1, deep-sea communities, for example, are coming to be estimated as far more diverse than supposed even a very few years ago: Thorson (1971) estimated 160 000 marine species, but Grassle and Maciolek (1992) commented 'As more of the deep sea is sampled, the number of species will certainly be greater than one million and may exceed 10 million'. In both marine and terrestrial systems, our current most-informed estimates of diversity may be wildly inaccurate. The problem for practical conservationists is to interpret such unknowns as effectively as possible with extremely limited resources, and to communicate with (and educate) people effectively to emphasize the importance of invertebrates in natural systems. This chapter discusses possible ways through the morass of uncertainties and impediments to understanding, which threaten to thwart efforts for invertebrate conservation.

With continued pressures for assessment of global and more local diversity, the factors that affect it, and for rapid surveys based on sampling sets which gather large numbers of specimens of ecologically informative taxa, the need for effective analysis and transfer of information is obvious. This alone is only initial documentation to be employed in practical conservation. The basic need subsequently is to communicate the knowledge gained to predict effects of change and to interpret this in changes manifest in the communities present, leading to management to counter threats and sustain taxa, communities, and ecosystems. An allied problem is that of retrospective analysis of samples that have not been preserved or documented optimally. Sorting, and deposition of voucher material (p. 150) in a recognized institution where it will be housed permanently and be available for future study, is highly desirable. This procedure is recommended as mandatory in some survey protocols, but is often disregarded or given very low priority because of logistic constraints. Simply 'recognizing' the species consistently depends on access to such vouchers, or to the large reference collections available in major museums and other research institutions. And the high proportions of undescribed species (for example, Hawksworth and Mound (1991), suggest that perhaps only around 3 per cent of the world's nematodes are described) emphasize the need for massively increased taxonomic expertise in order to interpret invertebrate diversity adequately. However, the current practicality is that numbers of taxonomists are declining. Despite the high political exposure of 'biodiversity', this prominence is not generally accompanied by funding benefits for such apparently routine tasks as identifying and describing its constituent elements as a basis for informed understanding. In short, as the need for invertebrate surveys and the effective interpretation of these is appreciated increasingly as a central theme underpinning the study and understanding of biodiversity, the global capability to do this is being reduced effectively, much of the existing corpus of expertise is being lost, and it is not being replaced. Systematics is not an attractive career option for excellent young biologists, and teaching of systematics is tending to be supplanted by more 'glamorous' aspects of biology.

The central roles of major museums as 'storehouses' of biological information useful in conservation assessment has been stressed repeatedly in recent years and, indeed, some of the most stimulating contributions to understanding invertebrate biodiversity have come from museum-based zoologists. But it seems to be expected that the scientific community can work interpretative miracles with little logistic support. They can not. Referring to biomonitoring by using benthic invertebrates, Brinkhurst (1993) wrote 'The search for a Holy Grail index that can be determined by the underfunded without exerting their minds persists, and is as futile now as it was in King Arthur's Britain.' Likewise, complex collections of material cannot be analysed without commitment, and there is no adequate substitute for this. Much of this chapter, however, addresses possible shortcuts

and varying levels of approximation to facilitate practical conservation.

It is often easy to recognize different species in samples as such, but the need for reliable and consistent recognition of 'morphospecies' becomes strongly apparent when one has to correlate series of, perhaps, hundreds of different samples. This can only be done with the aid of detailed comparative notes, in essence diagnoses, of the various forms, and continual comparison with voucher material.

In view of such difficulties, we might ask whether direct analysis of invertebrate assemblages has any value in their conservation, or whether invertebrates might be protected adequately under the more usual umbrella taxa of vertebrates and vascular plants. If this were so, invertebrates could indeed be omitted from direct assessment in such contexts—though this would not, of course, obviate the need to manage rare species if they were perceived to be severely threatened. Few studies have specifically examined this proposal. In the mallee area of western Victoria, Australia, Yen (1987) correlated the vertebrate and Coleoptera assemblages of 32 sites from eight vegetation communities. The beetles (total 176 species) used this habitat differently from the vertebrates (total 60 species), so that there was no significant relationship between the number of vertebrate species and number of beetle species found in each vegetation community, and ordination analysis revealed a greater number of discrete assemblages of beetles than of vertebrates. Many of the beetles were very restricted in distribution, and an area reserved on the basis of a rich vertebrate fauna may not necessarily be one optimal for beetles. The principle emerging is of much more general relevance; that many ecologically sensitive invertebrates are indeed very responsive to changes in the environment, and partition their resources and microhabitats in extraordinarily subtle ways, perhaps far more subtly than do most vertebrates. In providing for their needs, and for using animals to monitor environmental health, it is unrealistic to ignore the most abundant, diverse, and sensitive taxa in species assemblages. Simply, assessment of biotic assemblages at other than the most superficial level must include representative invertebrate groups, and the practical problems then devolve to selecting these and interpreting their diversity and abundance as effectively as possible.

Systematic interpretation

Logistic constraints emphasize further the need for the best possible focus in invertebrate conservation and the futility of inventory or complete documentation, however desirable this may be as an idealistic goal. For insects alone, conservatively estimating a diversity of 10 million species, current levels of global commitment to systematic study would not be able to describe the world fauna in less than 1000 years (Mound and Gaston 1994). As Hawksworth and Mound (1991) concluded, 'These are global problems requiring international co-ordination and effort on a level hitherto not contemplated.' Even for a select range of 'priority' invertebrate groups (such as major indicator taxa and keystone species, for example) preparation of the necessary illustrated keys and manuals to facilitate recognition and identification by non-specialists, association of early stages with adults, and estimates of regional diversity patterns, are usually far off. The value of ants as an indicator group was noted earlier (p. 32). The recent encyclopaedic volume by Hölldobler and Wilson (1990), enables us to recognize most ants to genus level, but much of their value as indicators and community monitors is at the species level, which is far more intractable. Notwithstanding the considerable importance of ants recognized to the 'functional' level or to genus, the species of ants in most parts of the world can be differentiated consistently by only a handful of specialists, and many of them are yet un-named. One major value of comprehensive voucher collections, therefore, is as a repository of unpublished (or unpublishable) information on biodiversity. If the ants from a survey are deposited thus, they can then be examined by other workers in the context of related work in the same or different areas. Taylor (1983) was one of the pioneer advocates of 'voucher numbering systems', by which reference can be made to taxa without formal names in a consistent and meaningful way: for many years,

his curation of ants in the Australian National Insect Collection involved giving appellations of a genus–number combination (such as 'Pheidole sp. No. 60'; 'Iridomyrmex sp. No. 12'), so that the collection became of considerable value as a reliable database even on taxa lacking conventional binomial names. This lack does not in any way preclude accumulation and communication of biological information. Key's and White's seminal studies of evolution and genetics of morabine grasshoppers (for overview and references, see Key 1982) is one classic Australian example in which considerable understanding, of great value in conservation through helping to define the taxonomic and distributional limits of species of a vulnerable endemic group of insects, was obtained from numbered, rather than named, species. The eventual naming has drawn on this mass of additional insight to render the systematics of the grasshoppers stable and reliable on far more than morphological features alone.

Such formalized codes can be published in ecological accounts in the secure knowledge that they are associated permanently with specimens available for reference by others. This level of collection-based 'species with code identities' thus permits data on the species to be indexed, analysed, retrieved systematically, and reported before the formal nomenclature is established.

The taxonomic process (Fig. 8.1) sets out to transfer 'species in nature' via 'species represented in collections' to 'species taxonomically understood', and each step is subject to many variables (positive or negative multipliers) (Taylor 1983).

The 'accession rate' is labour intensive in that it reflects collecting intensity and effort, the amount of attention given to fieldwork, and where this occurs. In many temperate regions, additional collecting will add relatively low proportions of 'new' species to the collection, whereas in the tropics the percentage rises dramatically, especially if novel collecting techniques are used or poorly known groups are targeted. The specimens are not represented properly in collections until they have been preserved, prepared, and labelled, a process that depends on technical and curatorial support and which can be expensive. Proper conversion of raw samples to the state where they are ready to be sent to specialists can lead to a ratio of sampling costs to processing costs as low as 1 : 40 for terrestrial arthropods (Biological Survey of Canada 1994). The 'naming rate' (Taylor 1983) reflects taxonomic activity—not only the number of taxonomists (and few institutions employ more than one or two for any large group of invertebrates; many have no specialists on the majority of phyla; and 'the entomologist' may specialize on one order or family of insects), but also that most such people can spend only a proportion of their time on primary research. The final synthesis brings the various formal descriptions together into overviews and reviews, by which the evolution of the group is clarified, the relationships between various constituent taxa

suggested, and the phylogeny advanced so that additional species can be placed accurately in the systematic framework established.

The level 'species in nature' varies in response to evolution (+) and extinction (−), and the systematist's goal is to transform this to 'species taxonomically understood' (Fig. 8.1b). As taxonomic understanding increases, concepts of the limits of species within a group become refined and the application of names may change. One great value of voucher material is that it is available for reappraisal and checking if this is needed. Voucher specimens 'physically and permanently document data in an archival report by (1) verifying the identity of the organism(s) used in the study and (2) by so doing ensure that a study which otherwise could not be repeated can be accurately reviewed or reassessed' (Lee *et al.* 1982). These authors suggested that voucher material should be required and recognized as an integral budgeting component of any survey, a message of considerable practical relevance in invertebrate surveys.

Employment of biological diversity technicians to collect or analyse samples of invertebrates is often advocated nowadays to help meet the increasing demands for taxonomic services. However, any faulty estimation of the numbers and identities of species may have far-reaching consequences and impede, rather than enhance, understanding, so that great care is needed to ensure the quality of the results produced. In trials of the efficiency of performance of non-specialist workers in assessing samples of aquatic insects, Cranston and Hillman (1992) found variability associated with:

(1) small body size of the specimens;

(2) increase in number of closely related taxa (when lower-level distinctions were difficult); and

(3) morphological variability within species (such as the larval instars of some damselflies, Zygoptera).

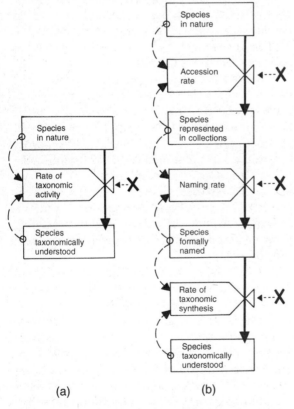

(a)   (b)

Fig. 8.1 The taxonomic process (Taylor 1983). (a) The basic flow of 'species in nature' to 'species taxonomically understood' is influenced by the rate of taxonomic activity; (b) the various steps in this sequence are each open to variability because of the amount of logistic support and priority given.

Such problems are likely to be common to many other invertebrate taxa, perhaps especially immature stages, and emphasize the need for substantial specialized training. The assumption that non-specialist personnel can replace specialist taxonomic expertise at a fraction of the cost is patently false, but there is a major role for such technicians in extending the efficiency of the limited number of taxonomists by effective sorting and preparation of specimens from bulk samples, and thus rendering specialists' time more effective, as has been pursued through INBio in Costa Rica (p. 10) to help document and assess the invertebrate fauna of that country. For benthic invertebrates, Brinkhurst (1993) found that it was cheaper to use the best taxonomists directly rather than just having them verify specimens identified by relatively unskilled staff. However, any practical shortcuts towards rapid biodiversity assessment need to be incorporated wherever they can be made reliable, and one criterion for focusing on particular groups for enumeration could well be the ease with which suites of species can be recognized by non-professionals with appropriate training. For now, and for the foreseeable future, the 'parataxonomist approach' relies on the international pool of taxonomic specialists, rather than any substantial indigenous intellectual capability (Mound and Gaston 1994).

Ecological interpretation

Effective assessment of invertebrate biodiversity thus has a number of components or phases:

(1) seeking possible shortcuts by concentrating on particular useful or informative taxa which mirror adequately the natural complexity of the communities they represent, through

(2) sampling and processing the material, and

(3) understanding its variation in time and space and how it is influenced by human activity, to

(4) developing systems for recording, analysing, retrieving, and communicating that information rapidly and efficiently, perhaps in combination with other relevant information.

For example, the coincidence of particular vegetation types or plant species, or of soil type, with particular kinds of terrestrial invertebrates may give valuable leads to understanding the factors influencing natural distribution, and plotting land-use patterns with historical and recent distribution may reveal factors associated with decline or expansion, and possible threats. This approach is in its infancy for most of the tropics, in particular. For some temperate-region countries, though, effective databases have been set up, and are augmented continually. They usually commence by the 'simple' mapping of selected groups, as discussed on p. 126, in some form of geographical 'grid' form, but can then be elaborated to constitute the basis of geographical information systems (GISs) of value in detection of critical faunas, centres of diversity, refuges, and, in short, for setting objective patterns of priorities in conservation, limited only by the extent and reliability of the data incorporated.

Three approaches may alleviate the need for complete species-level inventory:

(1) selection of indicator taxa or other 'priority groups';

(2) taxonomic reduction; and

(3) allocation to functional groups (Cranston 1990).

The first of these has been discussed earlier (p. 25). The second involves reducing the level of identification of samples, either to groups above species (such as genera or families) or to 'morphospecies' or similar 'recognizable taxonomic units' (RTUs). Use of the latter approach depends on the responsible deposition of voucher material, which becomes the only means by which workers can identify that material in the absence of published binomial names. Under some circumstances this approach can be very effective (p. 149), but there is a danger that uncritical or incompatible voucher numbers may proliferate as an easier option to more critical appraisal, and out-distance taxonomic capability (Cranston 1990). The third approach involves allocation of species to functional guilds by reference to knowledge of their biology: it is almost inevitable that approximations, possibly misleading, will result because of extrapolation of limited information (usually based on very few species) to others. Many taxonomic groups reveal surprising (or, at least, unexpected) ecological variety among their species when they are examined in detail. Recognition of ecological groupings could be used to discard need of taxonomic recognition, as redundant, but this substitution cannot be justified at present (Cranston 1990). The assumption that all (or most) members of a taxonomic group, such as an insect family, behave in a uniform way is difficult to validate and is most commonly simply not true.

Sampling methods to estimate species richness of any defined areas must be as efficient as possible. The major needs (Coddington *et al.* 1991) are that they should be (1) fast—because time is often restricted and expensive; (2) reliable—because they may need to be used in various areas to produce comparable data; (3) simple and (4) cheap—the last two because they will need to be used in many places where infrastructure is basic and complex apparatus is not available. Development of pragmatic tools and protocols for rapid or reliable sampling efficiency is vital. In a thoughtful appraisal of traditional approaches to estimating biodiversity (that is, by ecologists and by systematists: the former may have long-term sampling programmes and be 'plot-based' and the latter may visit an area once or infrequently and sample more extensively), Coddington *et al.* suggested that the following are important criteria for sampling protocols.

1. The usual complement of collecting techniques used by museum personnel involved in inventories of particular groups should be modified as little as possible in order to yield analytically tractable data.

2. The number of collecting methods should be as few as possible and sampling protocol should also be simplified by using a 'simple' classification of microhabitats. Methods should be chosen for their efficiency and low overlap with other mehtods, ensuring maximum complementarity.

3. Protocols should work well in plot-based and plotless sampling situations. 'Time spent sampling' may be a useful index of sampling intensity.

4. The sample unit should be sufficiently large to yield adequate numbers (individuals and/or species) but sufficiently small that the number of samples taken is adequate for statistical comparison.

5. Data should be collected so that variation can be estimated and analysed, in relation to factors such as site, season, sampling method, time of day, and so on.

6. Data on numbers of individuals and species should be able to be combined to produce species abundance distributions usable to estimate species richness.

7. If possible, the analytical approach should yield confidence intervals on the estimates.

Invertebrates and priority sites

The kinds of data needed for ecological assessment and ranking, and its applications, were described by Pellew (1991), who stressed the dependence on the knowledge of local people for a pragmatic approach to selecting areas of high biological value over much of the tropics. In tropical forest regions of Central Africa, for example, Pellew noted that 104 areas had been selected as 'significant' on the criteria of:

(1) apparent or documented species richness;

(2) presence of rare species;

(3) site representative of the region;

(4) site including special communities or habitats of restricted distribution (such as montane forest);

(5) site of high apparent species richness but threatened by some human disturbance; and

(6) important for flagship species or for wild relatives of crop plants.

The 'most available' knowledge from indigenous people in these contexts is for vertebrates—particularly mammals and birds—and the approach emphasizes (yet again) the lack of realistic ability

Table 8.1 Threatened invertebrate communities designated by Wells *et al.* (1983)

Locality	Threat(s)	Significance
Usambara Mountains, Tanzania	High human intrusion Deforestation	Many endemic taxa: insects, diplopods, molluscs Outstandingly rich community Centre of evolution; important research site
Gunung Mulu rainforest, Sarawak	Increasing pressure on surrounding areas	Major National Park of Sarawak, with almost all inland forest categories present Very high diversity; specialized cave communities
San Bruno Mountain, California, USA	Urbanization: weed invasions, recreation	Remnant mountain habitat in San Francisco (Number of rare or endemic insects
Banks Peninsula, New Zealand	Deforestation; urban development	Remnant forest areas; large number of endemic species; type locality for many others
El Segundo sand dunes, California, USA	Invasion of exotic plants Recreation	Isolated remnant ecosystem in Los Angeles, Unique dune communities; type locality for number of endemic or restricted species
Dead Sea depression, Israel, Jordan	Fragile systems Agriculture and more general 'development'	High invertebrate endemism, including freshwater taxa
Cueva los Chorros, Puerto Rico	Vulnerable to visitor activity: small	Very rich invertebrate fauna
Deadhorse Cave, Washington, USA	Vulnerable to surface logging, and fouling of stream	Diverse stream and terrestrial fauna
Sumava Mountain mires, Czechoslovakia	Mainly low threats: some possible changes	Relict peat bog habitat for many restricted/threatened insects
Taka Bone Rate coral atoll, Indonesia	Overfishing, including use of explosives	Largest coral atoll in Indonesia: high diversity—including 6 spp. of giant clams
Roseland marine conservation area, UK	Dredging activities for maerl (red calcareous algae)	Richest marine habitat in British Isles, largely associated with maerl

to promote invertebrate parameters over the parts of the world where they have greatest relevance and are most diverse. Much inference on the worth of particular tropical sites for invertebrates derives from knowledge of expatriate naturalists who may have lived in the area. It is therefore fortuitous rather than planned.

One way of adjusting this imbalance was exemplified by Wells *et al.* (1983) in their designation of 'threatened communities' of invertebrates, in addition to single taxa, with equivalent red data book rankings. The examples are ecologically and geographically widespread (Table 8.1) and represent only a few of the communities that could be advanced as important. They encompass many of the 'species-based parameters' used to rank sites or communities (p. 86), but approaches to setting global priority areas (Williams *et al.* 1991) and critical faunas (p. 79) mark an objective route to defining such communities more satisfactorily.

Modern conservation biology can be made more effective by attempting to improve understanding of several themes, discussed by Soulé and Kohm (1989) (see also Holdgate 1991). They include:

(1) global patterns of distribution of biological diversity;

(2) the functioning of communities, especially the interactions that maintain them, and the relevant time scales;

(3) the extent to which communities can tolerate disturbances;

(4) the effects of different kinds and amounts of disturbance;

(5) the consequences of fragmentation;

(6) the effects of species introduction, and

(7) the reproduction and propagation of selected species.

Collectively, these would enhance ecological understanding immensely, but the likelihood of comprehensive documentation is indeed small.

The 'Sustainable Biosphere Initiative' (Lubchenko et al. 1991) also stressed that basic research to acquire ecological knowledge is essential in the rational management of Earth's resources, and that this must be communicated and utilized effectively. They suggested, as priorities for biodiversity research:

(1) biological inventory;

(2) the biology of rare and declining species; and

(3) the effects of global and regional change on biological diversity.

All, stated so blandly, are daunting.

These are a basis for developing guidelines and well-funded research programmes for conservation and management, including restoration. Any of these is relevant to invertebrate conservation, but the magnitude of the task is indeed daunting. In order to make the most of limited resources, further objective priorities need to be established, and a widespread education programme to reduce public prejudice against invertebrates and to foster appreciation of their positive values at all levels is needed.

The diversity of invertebrate life-forms, and our lack of knowledge of most of the phyla, hamper assessment of their use and needs in conservation, and necessitates a pragmatic approach to advancing invertebrate conservation awareness and action. Planning necessitates appreciation of scale, a lesson that emerges repeatedly as the needs for management of particular habitats, sites or taxa are refined progressively. Samways (1994a) has stressed the conceptual value of appreciating scale differences at the levels of host ('plantscape') to 'landscape' in planning insect conservation, and the need to assure the different resources needed by various disparate life stages such as caterpillar and butterfly. What an insect, snail, or crab 'sees' as its main requirements may differ markedly from human perception.

PRIORITY TAXA FOR ASSESSING NEED

With limited resources some form of 'triage', or concentration of effort on selected taxa, to advance effective invertebrate conservation seems largely inevitable, however ethically abhorrent this may be to many people. Perhaps the surest way to assure the well-being of the greatest number of invertebrates is to ignore many groups almost entirely—surely a puzzling priority advocation! This is not, in any way, equivalent to condoning their loss, diminishing their importance, or suggesting that they are 'expendable'. They are not; many play vital ecological roles and their loss would be far-reaching and tragic, from many points of view. It is, however, pragmatic. With such restricted resources available to conserve and assess biodiversity, attempts to target specifically groups about which we know very little may dilute the capability to conserve invertebrates as a whole and result in lack of understanding of all groups.

Although advocacy of neglect may smack of the traditional omission of invertebrates from many conservation agendas, it is very different. If a suitable range of invertebrate groups could be selected by consensus to receive the bulk of our restricted available attention for status evaluation and conservation assessment, and the greatest share of the limited 'invertebrate conservation dollar', many—perhaps most—of the others might be conserved effectively under this 'extended umbrella', far more effectively than any invertebrates could be sheltered by knowledge of vertebrates or plants alone, *or* by a more superficial understanding of a wider array of invertebrate taxa. This does not, of course, preclude any study of notable taxa in the de-emphasized groups. But 'priority' must imply that something else is not done, that other groups are not studied in equivalent detail.

The preferred groups could be selected on ecological and other values, and the following suite of features incorporates many of these criteria for an initial appraisal of 'priority phyla' for accumulating a focused conservation database (New 1993*b*):

1. Include marine, freshwater, and terrestrial taxa, and groups combining these in various combinations.

2. Ensure representation of diverse ways of life, with all major feeding modes and trophic levels replicated extensively.

3. Use groups that are geographically widespread, but which also include local endemics, known 'critical faunas', or known centres of diversity in a collective range.

4. The groups should be accessible and amenable to quantitative sampling by standard (or near-standard) techniques.

5. The groups should be reasonably diverse but with an established taxonomic framework for all or significant relevant parts.

6. Preference should be given to groups for which substantial ecological information exists. The preferred groups noted here include virtually all invertebrate taxa that have been promoted as indicators and the like, and a high proportion of 'species-focus' conservation cases. For many, there is thus a predisposition to effective conservation through knowledge or history of responses to disturbance, so that existing foci can be used as bases from which to expand effort systematically. Within each large group, indicator segregates can be identified progressively.

7. 'Values' for most are defined or definable, at least in general terms. These include a number of 'commodity values' likely to engender political sympathy and support.

8. Public prejudice may be minimal for most.

9. For each group, a 'critical mass' of specialist scientists should exist, including people working in geographically varied areas. There is thus potential for co-ordinating and designing broad-based sampling regimes, identification programmes, documentation of values, and so on.

If taxon-focusing was pursued at this major taxonomic group level of separation, not more than eight or nine phyla, and only parts of these, would satisfy the majority of these criteria. Compare these (Table 8.2) with most others, many of which proffer severe barriers to rapid practical incorporation into conservation strategies on grounds such as:

(1) small size;

(2) inaccessibility, and difficulty of sampling;

(3) lack of taxonomic knowledge and agreement;

(4) lack of ecological knowledge;

(5) overwhelmingly large, or highly uncertain, levels of diversity;

(6) an entirely parasitic mode of life, so that existence is entirely dependent on that of more accessible organisms; and

(7) high level of public prejudice through 'causing harm' or mere ignorance.

Each of these barriers is difficult to overcome; collectively they are insuperable with present or foreseeable logistic support. These de-emphasized groups are therefore those for which

(1) underlying levels of relevant knowledge are lowest;

Table 8.2 Invertebrate groups useful in conservation planning, and which might be major foci for documenting invertebrate assemblages

Phylum	Habitat	Reason
Cnidaria	Aquatic	Ecological importance, commodity
Porifera	Aquatic	Pollution, commodity
Platyhelminthes (Turbellaria)	Freshwater	Pollution indicators
Mollusca	Aquatic, terrestrial	Indicators, commodity, ecological importance, variety
Annelida	Aquatic, terrestrial	Pollution indicators, ecological importance, commodity
Onychophora	Terrestrial	Umbrellas in forests
Arthropoda (most)	(All)	(Many)
Bryozoa	Aquatic	Pollution indicators
Echinodermata	Marine	Ecological importance, commodity

(2) ecological understanding is grossly inadequate;

(3) the taxonomic impediment is high and unlikely to be overcome;

(4) the record of conservation interest is least; and

(5) public goodwill is least.

Each of these groups of animals does, of course, have devotees, and any specific taxon-focused conservation need should be addressed; but their higher-level role in developing any concerted strategy for invertebrate conservation as a whole is likely to be minimal. Even though, for example, some soil nematode faunas may reflect community maturity (Bongers 1990; Yeates 1994), interpretation has so far been only at the family level, with unknown ecological nuances within each of those large groups.

Within the 'preferred phyla' listed in Table 8.2, considerable further focus is needed, on pragmatic and ecological grounds, and to include as many of the 'selection criteria' as possible. Some groups of arthropods are clearly more predisposed to constructive assessment for conservation than others, for example. A different approach, but with similar emphasis in streamlining assessment and reducing the number of groups for which detailed systematic appraisal would be needed for assemblages would be to select optimal groups at a lower taxonomic level, perhaps of orders or families, over a broader range of phyla. For each preferred group, a series of target priorities is needed to guide the framework of inventory, assessing and monitoring diversity, and distribution, and understanding the conservation needs of major taxonomic components. Likewise, the optimal groups may differ for different habitats.

On a somewhat broader basis, Di Castri *et al*. (1992) recommended the following ideal representation of taxon sets for biodiversity investigation:

(1) all major functional guilds;

(2) the full range of growth forms for vascular plants and fungi, and the full range of body sizes and forms for animals;

(3) any keystone species or representatives of keystone groups;

(4) a full range of geographical distribution characteristics, from narrow endemics to cosmopolitan species; and

(5) representatives of species-rich and species-poor groups, relatively abundant groups, and relatively rare groups.

They emphasized the need to include selected groups of insects and fungi because these represent roles and sizes not measured by more easily studied groups.

Having selected a set of target groups in some

way, protocols and pro formas for their study can be developed and co-ordinated. As umbrella taxa, their collective diversity and abundance is assumed to reflect that of less well-understood invertebrate groups, so that estimates of patterns of biodiversity and the factors influencing it become increasingly relevant. This assumption itself is open to question. Cranston and Hillman (1992) found that the diversity of any particular group may not represent community diversity; but they and others have stressed the need for baseline data to clarify this. Such baselines for a number of coexisting groups are likely to be particularly valuable. Kremen *et al.* (1993) stressed the great value of terrestrial arthropod assemblages as a data source in conservation planning, because of the diverse contexts in which they occur and the great variety of animals involved. Following Di Castri *et al.*'s recommendations (1992), the following objectives are relevant for any given group, and these can be 'overlain' in various ways to examine possible determinants of significance for conservation management. The overall objectives include:

1. To use the study of given groups to estimate the total diversity of a given area, by sampling intensively at one or more sites.

2. To determine the influence of biogeographical factors on diversity by comparing different areas having the same ecosystems.

3. To compare sites representing different ecosystem types by intensive sampling to estimate influences of ecological factors.

4. To establish a monitoring system to identify the effects of various disturbances or potential or actual threatening processes.

5. To evaluate the biodiversity of any particular trophic group or level.

Together, these parameters contribute towards an inventory of biodiversity in an ecologically logical way and, by integration of intensive and extensive sampling programmes (Solbrig 1991), allow development of a more generally applicable methodology. Di Castri *et al.* (1992) recognized the need to utilize different taxa in different ecosystems—in the invertebrates, they recommended selected families of spiders and 'about 10–15 selected families or subfamilies' of insects for joint use in tropical and temperate forests, and noted that the insect groups should include (but not be limited to) groups that are already reasonably well known. It would clearly be advantageous to include groups known to have a useful indicator role in any particular vegetation type, or across several of these, or other, habitats.

Whereas a high level of complementarity and potential to compare between habitats and sites in different places is desirable, some groups are likely to be of most use in conservation assessment on a regional (rather than a global) level. Thus, tardigrades and rotifers may both be useful groups for assessment of low-diversity Antarctic communities (Usher and Edwards 1986) but less so in most other parts of the world where they are less prominent and only poorly understood. Speight (1986*b*) suggested selecting a suite of 'foundation groups', to be sampled on all sites and additional 'auxiliary' groups to be sampled in particular environments. Thus in his example of seeking insect bio-indicators in Ireland, carabid beetles, hoverflies, and sawflies (Symphyta), each with about 600 European species, could be useful foundation groups, and additional families incorporated for wetlands, forest or woodlands, or open ground environments.

PRIORITY HABITATS

Directories of protected areas are gradually being produced by the World Conservation Union and these may be a constructive focus for priority for estimating invertebrate diversity. Especially in the tropics, many of these represent enclaves of 'relatively natural' communities. However, details of even well-known invertebrate groups present are usually unavailable and organized surveys of invertebrate groups to determine which taxa are present in reserved areas are an important need. The salvation of many of these areas may depend on tourism or other income-producing activities, but resource demands or political turmoil may prevent this and lead to loss of integrity of such areas in many places.

Table 8.3 IUCN categories and management objectives for protected areas (after IUCN 1990*b*)

Management category	Name	Aims
I	Scientific reserve/strict nature reserve	Protection for ecological representation and maintenance of genetic resources
II	National park	Protection of areas of national or international significance for scientific, educational, and recreational use
III	Natural monument/ Natural landmark	Protection of significant natural features
IV	Managed nature reserve/wildlife sanctuary	Protection of significant taxa or communities where specific human manipulation is needed to maintain natural conditions
V	Protected landscape or seascape	Maintenance of significant landscape while providing for recreation and tourism for public enjoyment
VI	Resource reserve	Protection for future use, and prevention of exploitation without appropriate planning and objectives
VII	Natural biotic area/ anthropological reserve	Protection of way of life of societies from disturbance by modern technology
VIII	Multiple use — management area	Provide for sustainable production of resource (e.g. water, timber) orientated to support of economic activities

The higher categories emphasize biodiversity values and the lower categories favour sustainable use values.

A representative *Directory of South Asian protected areas* (IUCN 1990*b*), exemplifies many of the management problems involved: in essence the traditional conservation approach of 'locking up' areas and protecting them from change is becoming extraordinarily difficult to achieve as human populations increase, and this aim is commonly being superseded by approaches involving management to support sustainable development and allowing changes of various kinds. A range of management options has been defined for protected areas (Table 8.3).

In addition to the categories listed in the table, areas in category IX (biosphere reserves) have the role of integrating conservation with provision for an international network of areas for research and monitoring, and for sustainable use of their natural resources. This involves zoning reserves in various ways—'sacrosanct' core areas, buffer zones, traditional use zones, and so on. Lastly, category X areas (world heritage sites) protect natural features considered to be of world heritage quality. A fuller account of these categories is included in McNeely and Miller (1984).

In addition, the approach pioneered in Britain and elsewhere in Europe of an 'invertebrate

site register' may be feasible in other regions, though perhaps not over much of the tropics. This entails listing and designating sites known to support taxa of conservation interest or significance, as a first step in selecting and documenting regional needs at this level. As with designating priority species, it is vital to maintain credibility in any exercise of this nature.

From the biodiversity conservation viewpoint, the notion of taxic complementarity (Vane-Wright *et al.* 1991, Williams *et al.* 1991), that is of using the number of additional taxa not represented in existing reserves as a guide for giving priority to additional sites, is a useful guiding principle, as emphasized in Chapter 4. For invertebrates, though, we rarely (if ever) have definitive listings of species of any group already present, except in some temperate regions. However, concentration on selected groups of invertebrates (particularly from among the 'preferred groups') is likely to be generally useful in augmenting knowledge of the ecological worth of existing protected areas and understanding what their future priority needs for conservation may be. Many early protected areas received little scientific input into their design, but were established for aesthetic or socio-economic reasons

(Leader-Williams *et al.* 1990). Sound scientific rationale for designation of priority areas is now more relevant and important than ever before, as conflicting priorities for land use manifest with increasing frequency and strength. The need for sound management of the world's remaining near-pristine areas, to assure their sustainability, is vital: without incorporating consideration of the major groups of animals, and assuring that the knowledge available is adequate for at least basic understanding of how these contribute to the integrated functioning and worth of natural communities, it is difficult to see how any such management can be wholly satisfactory.

GLOBAL CHANGE

However diligently and competently we may apply scientific principles and knowledge to conservation of natural communities and successfully counter most threats to them, there remains one major threat—perhaps the dominant conservation issue facing us—that of artificially increased and rapid climatic change reflected in global warming, the 'greenhouse effect'. Despite continuing debate over the existence, rate, and extent of global warming, many scientists accept its reality (see papers in Pearman 1988) and that conservation planning should incorporate likely scenarios resulting from this form of stress. Peters and Darling (1985) emphasized that, if the greenhouse effect eventuates, it 'will pose a new and major threat to species within reserves, species already stressed by the effects of habitat fragmentation'. Parsons (1989) noted that extinction and replacement of rare tropical rainforest *Drosophila* flies in Australia is likely to occur if temperatures increase by as little as 2 °C. Progressive modelling of the tolerances of many other forest invertebrates is likely to reveal many parallel instances.

Likewise, many corals are at present near their limits of thermal tolerance, and the increase in frequency and extent of 'bleaching' of corals noted recently may indeed reflect global warming. 'Bleaching' is the loss of symbiotic zooxanthellae or reduction in their photosynthetic pigments, and its causes are not understood fully (Jackson 1991; Glynn 1993), although temperature rises of 1–2 °C have been suggested as a factor contributing to this mortality.

The frequency, scale, and severity of coral bleaching reported recently is unprecedented. In many cases, coral reef bleaching was reported to occur during the summer or towards the end of a lengthy warming period, and under calm conditions which favoured localized heating and high penetration of ultraviolet radiation. The effects of large-scale coral bleaching on the numerous other animals associated, often obligately, with reefs is also a major concern, but has barely been investigated until now (Glynn 1993).

Climatic tolerance limits of an invertebrate (or other animal or plant) species are the main determinants of its 'fundamental niche' (the set of environmental conditions under which it could live: Hutchinson 1957), with the 'realized niche' (its actual regime) constrained by interactions with other taxa. Changes in tolerance limits have been reported in species introduced into temperate regions from warmer areas and which depended initially on heated conditions. The freshwater oligochaete worm *Branchiura sowerbyi*, from South-East Asia, occurred in heated aquaria and warm outflows from power stations in Britain after it was introduced, but has more recently been recorded in unheated river sites (Aston 1968; Macan 1974). Several freshwater snails have shown similar adaptations in Britain, and temperature may be particularly important to many freshwater organisms. Examples such as the adonis blue butterfly (p. 97) and the Roman snail (p. 136) indicate the dependence of terrestrial taxa on small habitats with favourable temperature regimes on the fringes of their geographical range. Much relevant background information on such adaptations is included in Ford (1982), who noted that the ranges of migratory species and patterns of disease infestation of invertebrates are among the many other factors likely to be influenced by changing climates.

Global warming is likely to increase the susceptibility to extinction of many invertebrates which are unable to adapt to warmer regimes, because of either the rate or extent of change, or the resultant loss of resources (such as foodplants) on which they depend.

Table 8.4 Kinds of community likely to be affected particularly by global warming over the next century (Peters and Darling 1985)

Community	Rationale
Peripheral populations	Near edge of contracting range
Geographically localized species	Small initial range
Genetically impoverished species	May lack flexibility to adapt to change
Specialized species	May be less tolerant of ecological change
Poor dispersers	Lack ability to move to suitable areas
Annuals	Possibility of complete reproductive failure in a given year, leading to extinction
Montane and alpine communities	Shift in species distribution in extreme environments
Arctic communities	Temperatures may increase relatively more than at low latitudes and provide correspondingly higher stress
Coastal communities	Possible changes in sea level leading to shift of communities

The land snail, *Arianta arbustorum*, became extinct at 16 of 29 localities near Basel, Switzerland, between 1908 and 1991 (Baur and Baur 1993). Because the snail is a simultaneous hermaphrodite, it may not be as susceptible as many other species to population decline *per se* where reproduction would depend on individuals finding a mate. One factor correlated with local extinctions is local climatic warming due to increased thermal regulation from built-up areas. Remote sensing data showed that temperature of the vegetation in habitats where *Arianta* has gone extinct is significantly higher than in habitats where it has persisted. Experimental data demonstrated that the snail's eggs do not develop in sites where the temperature exceeds 25 °C, so that climate warming may here be reflected directly in curtailing the development of a species living near its thermal maximum regime.

Peters and Darling (1985) noted nine kinds of communities which might be affected particularly by warming during the twenty-first century. These are listed here to indicate the massive influences that could occur on distribution and existence of invertebrates, and in interactions reflecting community composition (Table 8.4). All of these are potentially significant.

Alpine or montane invertebrate assemblages, for example, are often highly characteristic as they support numerous specialized species that

do not occur elsewhere: they have been targets or priorities for conservation in many parts of the world. However, such attempts may prove futile if the scenario shown in Fig. 8.2, whereby high-altitude specialists are rendered progressively extinct by climatic change, eventuates. Such species, shifting upward in response to warming,

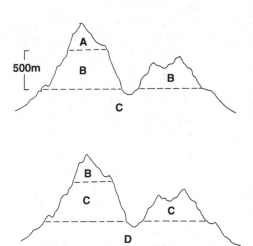

Fig. 8.2 Possible effects of global warming on distribution of alpine species. Top : 'present distribution' of species A, B, C. Bottom: distribution after 500 m shift in altitude in response to a 3° rise in temperature. A becomes locally extinct; B occupies decreased area, on fringe of range; C becomes fragmented and more restricted; D colonizes lowest levels. (After Peters and Darling 1985.)

(1) come to occupy smaller areas, which may eventually disappear;

(2) have smaller and more localised populations, and thus

(3) may become increasingly vulnerable also to other factors.

In addition, many montane invertebrate populations are very isolated, with little chance of influx from other populations of the same species, or of recolonization of a site where an extinction has eventuated, even if the area remains suitable.

Ramifications of such change for management for invertebrate conservation are complex. At the least, they emphasize the need for multiple and large reserves for communities likely to be threatened, to conserve the greatest possible range of conditions and to preserve opportunities for local adaptation within sites. A variety of topography and altitudinal variations could also be advantageous. In addition, Peters and Darling (1985) noted the wisdom of maintaining flexibility of reserve boundaries, so that these could be changed in the future, as local conditions change, and to ensure that they continue to function adequately for the protection of significant biota. Where possible, reserves should be planned to avoid the scenario depicted in Fig. 8.3: essentially, reserves to protect particular invertebrate species should not (unless otherwise unavoidable) be solely in regions susceptible to rapid change in suitability for those species through climatic change within the foreseeable future. Reserved populations near a margin where conditions are deteriorating would be more threatened than ones at the opposite range extreme.

However, the concomitant scenario of sea-level rise with increased global temperatures may not allow for such possibilities. Large areas of low-lying country are likely to be inundated in many parts of the world, with loss of many specialized habitats such as mangrove swamps and low sand dunes. Few people have commented on likely effects on invertebrates: for Britain, Foster (1991) noted that areas subject to inundation would encompass the combined distribution of most fenland insects included in

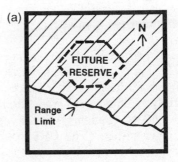

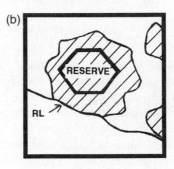

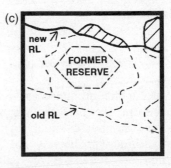

Fig. 8.3 Possible effects of global warming on integrity of nature reserves: (a) species distribution (shaded) before human habitation; (b) fragmented distribution ensues, with reserve designated to protect the species; (c) global warming results in contraction of species' range limit, so that the reserve is no longer effective. (After Peters and Darling 1985.)

the red data book, and relict fenland sites on slightly higher ground would become particularly significant as reserves for these species. Possible refugia against major effects of climate change need to be defined and these could become priorities for future reserve designation and enhancement.

Holloway and Stork (1991) made the intriguing suggestion that, as well as invertebrates being useful in indicating past climatic change, they may also be useful in monitoring effects of future changes. Samples of moth assemblages over 10 years on Norfolk Island showed responses to two unusually dry years by species being more evenly distributed. Particular species changed position in abundance rankings, so that study of changes in proportions of such taxa may reflect changing climate conditions over a longer period. The background information needed for more widespread investigation of this scenario elsewhere is currently sparse.

Moore (1971) commented 'If we knew the world was coming to an end in a few years' time, no-one would advocate conservation . . . conservationists . . . must take a calculated risk that their work will not be in vain.' The lesson must be to anticipate the future as well as we are able to, and incorporate the more predictable of likely changes constructively into conservation planning now, rather than waiting until that whole new suite of threats eventuates.

EDUCATION

Inclusion of 'public prejudice' as one of the 'selection criteria' for the foregoing list of priority taxa may be unwise—many of the invertebrates that many people do not like (spiders, worms) are otherwise of high priority because of their important ecological roles. It is at present pragmatic, because an important part of practical invertebrate conservation is that of gaining public sympathy and appreciation for the animals. Much of the needed change in attitudes must come through gaining the goodwill of young people. Yen (1994) stressed that zoos and museums may be highly important in fostering education about invertebrates to many sectors of the community. The traditional approach of museums showing dead specimens and zoos exhibiting live animals tends to impoverish both when invertebrates are involved. A combination of 'dead' and 'live' displays, utilizing models, dioramas, and video technology has potential to instruct and

excite visitors. Yen believed that such displays need not focus specifically on conservation but, rather, could target fundamental aspects of invertebrate importance, biology, and ecological roles, or relate to increased awareness of habitats with which many spectators are already familiar—such as gardens and ponds. Local concerns and notable features are a sure attraction, but the most important message to convey is 'understanding invertebrates' (Hancocks 1992). A number of institutions, including aquariums, do indeed have such displays; many are overshadowed by vertebrates, but 'butterfly houses' and the like give exhibition priority to particular invertebrates. In Australia, a 'giant earthworm' museum in Victoria is probably unique in focusing on an oligochaete of conservation significance. Some exhibitions are 'purely aesthetic' (or commercial) in their aim, and there is much room for improving their role in communication of information.

The promotion of invertebrates as 'pets' has considerable educational value. Whereas rearing caterpillars (of local taxa and, in some places, others, such as exotic silkmoths) has long been a widespread hobby, keeping of stick insects, spiders, giant cockroaches, and others is becoming more common, and the therapeutic value of touching invertebrates has been advanced.

The *1990 International Zoo Year book* (Olney and Ellis 1991) lists the major invertebrate taxa in collections of many of the world's major zoos and aquaria. Thirty-seven per cent (283/758) of these institutions list 'invertebrates'. On this figure alone, it would seem that invertebrates are indeed a major facet of zoological displays. However, a breakdown of the numbers of taxa listed (where this is possible: $n = 274$ institutions) (Fig. 8.4) shows that a very high proportion of collections include few species, and some may well be kept solely as food supplies for vertebrates. Indeed, 31 list only one invertebrate (and, in some cases, state 'one individual'), and many in the large second category appear to have few individuals. Only 26 institutions claim to have more than 100 invertebrate species and the largest total (270 species) is for Shirahama Aquarium, a specialist marine

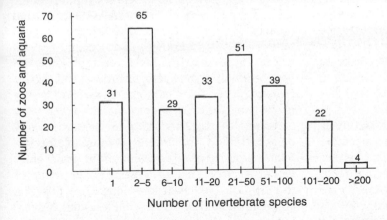

Fig. 8.4 Numbers of invertebrates maintained and displayed in zoos (data from Olney and Ellis 1991).

invertebrate display in Japan. World-wide, there is a tendency for aquaria to exhibit many more species of invertebrates than conventional zoos (Robinson 1991).

However, this lack of wide representation of many invertebrates does not reflect the substantial amounts of care and attention paid by a number of institutions to communicate ideas about invertebrates, to foster conservation awareness through rearing significant species, and in making notable advances in husbandry techniques for these and other taxa. For example, The London Zoo (with a large collection of around 120 species) is developing five breeding programmes for threatened invertebrates (Pearce-Kelly *et al*. 1991). Those for *Partula* snails and the Mexican red-kneed bird-eating spider were noted earlier (pp. 100, 101); the others are for the Olympia ground beetle, *Chrysocarabus olympiae* (p. 107), the rare bush-cricket known as the wartbiter (*Decticus verrucivorus*), and the robber crab, *Birgus latro*. The latter, known also as the coconut crab, is the world's largest terrestrial invertebrate, with individuals reaching to more than 4 kg. It is subject to considerable exploitation for food and tourist souvenirs, as well as to predation by introduced vertebrates, and habitat destruction. Despite their size and strength, *Birgus* appear to be susceptible to stress in captivity and are not easy to keep. The robber crab's requirements in captivity are not yet clear (Robinson 1991), but it may prove to be amenable to culture to supply commercial

demand in parts of the Pacific (Brown and Fielder 1991). However, slow growth rates and poor growth in captivity are likely to hamper such attempts: Vanuatu crabs take around 12 years to reach the minimum legal size for commercial use.

The London programmes are all joint undertakings with other institutions, reflecting a widespread policy of institutional co-operation for species of conservation significance.

Interpretative displays associated with live invertebrate exhibits are increasing in standard and appeal, as well as in the kinds of message they are designed to impart. Robinson (1991) noted six major characteristics of invertebrates that merit emphasis in educational programmes:

(1) diversity;

(2) abundance and biomass;

(3) complexity or radiation;

(4) history;

(5) biological and economic importance; and

(6) interactions with plants.

The need for combination of the traditional 'museum' and 'zoo' exhibition techniques is enhanced by the small size and cryptic habits of most invertebrates, and the measures by which such problems are being overcome are illustrated, as examples, by Yajima (1991, on the 'Insects Ecological Land' at Tama Zoo,

Japan) and Zabarauskas and Indiviglio (1991, on 'Jungle World' at Bronx Zoo, New York). Many current invertebrate displays concentrate on showing animals in conditions resembling their natural environments and thus differ markedly from their earlier counterparts, which tended to comprise a few sparsely labelled animals in cages, bearing little resemblance to nature.

The other main facet of education on invertebrates is the viewing and appreciation of the animals in their natural environments, either though 'field courses' or as a facet of ecotourism.

One example of an educational programme that promotes invertebrates is Sonoran Arthropod Studies Inc. (SASI), with the mission 'to foster an understanding of and appreciation for the arthropods and the vital roles they play in our natural environment and by so doing demonstrate the interdependence of all living things including humans . . . through publications, educational programs, research and exhibitions' (Prchal 1991). The organization has nearly a thousand members, produces a magazine *Backyard Bugwatching*, and operates a discovery centre in Arizona. About a thousand people participated in children's educational programmes in 1990, with another 35 presentations to 1280 students under Outreach Programs (SASI 1992).

Education is a more passive facet of other groups involved with invertebrate conservation, but the major thrust of those tends to be science. The major value of educating children, emphasized by Prchal (1991) is that they have not yet *learned* to 'hate bugs', and SASI also seeks to 'educate the educators' to this end.

A PRACTICAL AGENDA

This book has emphasized the differing nature of practical invertebrate conservation in temperate regions, especially in the northern hemisphere, and elsewhere. In Europe and North America, particularly, faunas are relatively small and well documented, there is a strong tradition of 'natural history interest' which has fostered a strong ethic for conservation, and museums,

universities, and learned societies support interest and expertise in invertebrate systematics and biology. By contrast, the more diverse tropical faunas are poorly known, and occur in regions where there may be little resident expertise or library and museum facilities, and where nature conservation may be perceived widely as an unaffordable luxury unless compelling economic or social benefits for people can be demonstrated. The major avenue of species-focusing pursued until now in temperate regions is not realistic for the future over most of the tropics. Yet such problems and differing perceptions cannot dismiss the fact that tropical faunas are the cradle of evolution of many invertebrate groups and that the practical conservation of many ancient and unusual lineages must seek to safeguard these as effectively as possible.

The following suggestions are effectively idealistic, but are advanced here to indicate some of the possible ways forward. Most of the steps are impracticable at present, but emphasize the need for two main thrusts:

(1) increasing education and effective communication as a means to increase understanding of invertebrates and resources to foster their conservation; and

(2) increasing scientific understanding and capability to translate these into practice.

This is not a simple 'recipe' but, rather, a series of points whose feasibility and role need to be considered actively and urgently as facets of practical conservation for invertebrtes.

Communicating the need and problems

1. Production of local versions of the European 'Charter for Invertebrates' (European Community 1986), as has been done for Australia (CONCOM 1989), emphasising the values of invertebrates by using local taxa as examples. This, or a similar document needs to be translated into local languages as necessary, and disseminated widely through schools, universities, and relevant goverment agencies. Where possible, its contents merit incorporation into

courses in sciences and/or conservation management.

2. Seeking subsidy for production of field guides to selected groups of invertebrates of use as flagships for conservation, or as priority taxa for rapid biodiversity assessment. For many countries, even broad recognition guides to major invertebrate groups (such as Harvey and Yen 1989) are not available or easily accessible. Likewise, primary taxonomic information may be available but only in expensive overseas journals or books; such information must be made available cheaply to those who can apply and augment it most efficiently, and avenues to do so locally usually do not exist. Lack of funding, anywhere in the world, for taxonomic work implies that support for such 'secondary' work may be very low—but it is of far greater importance for conservation management than completing documentation of many inconspicuous, hard-to-obtain, or poorly known taxa.

The levels of information needed vary considerably in different contexts. At one extreme, Onychophora have high evolutionary significance and rarity, so that the mere presence of any species may be notable (New 1994b) and the vital need is to recognize the general appearance of a velvetworm rather than to diagnose the species. We need to define other taxa for which this level of appraisal is adequate, rather than the other extreme of diagnosing particular subspecies of butterfly which can be, *per se*, sensitive targets for conservation measures. In general, such fine taxonomic resolution is difficult, sometimes arbitrary, and can be accomplished for few invertebrate groups. However, such relevant information must be made available, progressively, to non-specialists. An allied need is organizing the information that *is* available, in the form of summary bibliographies of useful keys and sources of information. Much of this tends to be 'taken for granted' by people with ready access to computerized bibliographic indices, but is simply not available elsewhere.

3. Providing for training to help people recognize the various invertebrate groups, as an extension of the above and to help improve local faunal knowledge. A major aim should be to establish a core of knowledge of local invertebrate faunas in many (ideally, all) countries. At this stage, this aim would necessitate funding for people in those countries to attend courses and workshops based in temperate areas: there are examples of these in Britain, the United States, Australia, and elsewhere, and there are encouraging signs that such opportunities are increasing.

4. Providing for training in rapid assessment of invertebrate faunas in marine (especially littoral), freshwater and terrestrial environments, and integrating this into standard conservation assessment protocols. The aims overlap with '3' but there is also need for a cheap general manual on relevant methods and theory, scientifically sound but couched in terms that can be applied by non-specialists. Again, several different language versions of this are needed, and it could provide the basis for comparative sampling programmes in many different regions and habitats.

5. Seeking subsidy for library purchases, including journal subscriptions, subscriptions to learned societies, and for participation of local scientists and managers in national and international forums on invertebrate biology and conservation management. As I write this, it is sobering to reflect that this book may never reach many countries in the less-developed world.

6. Wherever possible, increased access to reference collections, most of which are alienated from the country directly involved, is needed. At the least, replicate voucher collections of selected groups, preferably those for which accompanying identification guides are also available, need to be dispersed widely. This involves increasing resident/local curatorial and storage capability.

Applying and developing the science

1. It is inevitable that invertebrate conservation in the tropics must move away from high emphasis on rare species *per se*, simply because of the massive costs of such studies and the difficulties of even evaluating their status accurately. Single-species conservation programmes should be regarded as part of an

holistic effort to conserve natural habitats and ecosystems. Returns to practical conservation may be considerably greater by determining the patterns of diversity and distribution of selected, ecologically informative invertebrate groups and using those patterns in helping to set priorities for habitat conservation and effective protection of natural assemblages. There is thus an urgent need to define such priority invertebrates, and possible foci are those that may be used broadly as 'flagships', 'umbrellas' and 'indicator taxa'. Protocols for sampling and monitoring these then need to be defined and established, and resources obtained to do this. For virtually all protected areas in the tropics, there are no inventories or even partial listings of most invertebrate groups present. We usually have little idea of how effective those areas are in conserving invertebrates. Surveys of priority taxa in national parks and similar areas throughout the world are themselves a priority for establishing baseline information and guiding future management.

2. Integration of invertebrate conservation more firmly into sustainable habitat protection, and cultural and/or economic well-being. The invertebrates that are staple or luxury food items, or which have economic value in some way, are important locally in ways that most taxa are not, and such 'tangible benefit' is an important conduit for conservation if it can be harnessed effectively. 'Butterfly ranching' in Papua New Guinea (Orsak 1993) and elsewhere, for example, has helped many people to appreciate the value of natural habitat, and, by demonstrating that cash incomes can be obtained from enriching secondary habitats from the butterflies that breed there, is helping to reduce the rate of intrusions into remaining primary forest areas. Local support and sympathy for conservation is, of course, vital.

3. This example links with the major hopes for invertebrate conservation in the tropics—the abatement of major threats to habitats, and the effective protection of adequate natural habitats. The paramount threat remains that of habitat loss, and control of this may be expensive and politically difficult; it is important that the 'most valuable' areas are targeted for effective preservation urgently and, in many places, we are already reduced to considering remnant habitat patches as enclaves in highly modified landscapes. The needs are twofold:

1. To gain local and national support for habitat and biodiversity conservation in practical terms. The recent global 'Biodiversity Convention' (UNCED) is an important step in this direction, in gaining the support in principle of many nations in acknowledging that the future of the world's biota depends on massive action over the next few decades.

2. To gain practical means to pursue this.

For much of the tropics, some form of overseas aid is the only realistic avenue, and several recent cases indicate the kinds of intervention which are of massive importance for invertebrate conservation. As examples:

1. The United States government has recently (March 1994) (Anon 1994) committed US$20 million to help establish an Indonesian Biodiversity Foundation to be directed by an international board.

2. The Australian government, through the Australian International Development Assistance Bureau has (April 1994) pledged aid of A$4.27 million over the next 5 years to promote conservation of Queen Alexandra's Birdwing butterfly (p. 138) in the Oro Province of Papua New Guinea.

These operations represent, repectively:

(1) a potentially massive facilitation of the study of biodiversity in a diverse archipelago with enormous numbers of endemic invertebrates; and

(2) use of a notable flagship species to promote changes designed to protect habitats valuable for myriad less-conspicuous taxa.

Another approach, exemplified through INBio in Costa Rica (p. 10) is that of a major attempt to document tropical diversity being supported by a multinational pharmaceuticals firm in return for rights to screen organisms for any potentially useful biologically active compounds discovered, and a share of the income resulting from the development of these.

Lack of funding will remain the major impediment to practical conservation and, as Fiedler *et al.* (1993) commented, 'We must look at large scale solutions that encompass multiple approaches and fiscal wisdom in order to advance the goals of our global conservation efforts'. However, any avenues to practical sponsorship and funding should be explored. Several recent field explorations have been funded by naming new taxa after donors, for example. There may well be further opportunities in this area for selected 'attractive' invertebrate groups, with a generic patronym requiring a larger donation than a species' name. In this era of increasing environmental awareness, it is perhaps not too cynical to believe that industrial and development companies might welcome such tangible evidence of their concern for the places where they operate.

CONCLUDING COMMENT

The pioneering resolution on 'Conservation of insects and other invertebrates' adopted by the World Conservation Union in December 1990 (for full text and commentary, see Collins 1991) urged educational programmes of the kinds noted above, such as strengthened invertebrate displays linked (wherever feasible) to captive breeding and re-establishment programmes.

It also urged governments to:

(1) draft their national protective legislation recognizing that the primary threat to insects and other invertebrates is habitat destruction;

(2) broaden the scope and content of existing international conventions to make them more appropriate for insects, other invertebrates, and particularly their habitats; and

(3) promote practical recovery plans for invertebrate species already listed in national legislation and international conventions.

Governments and other agencies or organizations, as appropriate, were urged to pursue efforts to understand invertebrate biology and recognize the value of invertebrate conservation and the use of invertebrates in assessing conservation values, as well as to limit the incidence of major threats to them. Collectively, the points in the resolution recognize (1) the lacunae in our knowledge-base and the need to fill these, and (2) the need to counter threatening processes and ensure the safety of habitats for invertebrates. They stress that the major way forward in understanding the needs for invertebrates, and putting these into practice, depends on sound knowledge but also on the political support for implementing the science in practical contexts, and financial support for accumulating, integrating, and using the information needed for rational informed conservation. There is little doubt that, given adequate support, very good 'invertebrate conservation biology' can be done. However, without effective communication and the facility to influence policy, that work is incomplete in its most vital dimension: practical conservation.

Urging conservation of invertebrate animals is by no means new. Needs of various taxa have been promoted by bodies such as the Xerces Society (USA), the Joint Committee for the Conservation of British Insects (UK), Butterfly Conservation (UK), l'Organisation pour l'Information éco-entomologique (France), and the entomological societies and similar groups focusing on other invertebrate taxa, but those needs have accelerated in urgency and relevance during the past two decades. The World Conservation Union's Species Survival Commission includes several specialist groups focusing on particular invertebrate groups (molluscs, Lepidoptera, dragonflies, orthopteroid insects, waterbeetles, social insects) and others are likely to be formed. It has also initiated an invertebrate conservation task force to help guide the subject into the future. One result of this increased collective attention is the increasing profile of invertebrates in the wildlife protection legislation of many countries, and the proliferating lists of 'threatened invertebrates' in many parts of the world. Such high profiles demand responsible action. If particular butterflies, molluscs, or other invertebrates are to be accorded status equivalent to that of charismatic mammals or birds, as many conservationists promoting such listings would urge, they may well merit equivalent effort to assure their safety. However, although the loss of any species is regrettable, the major thrust of our efforts—that which will affect the environment of our descendants most directly—must be to slow, even curtail, the enormous rates and levels of extinctions likely to occur. Such measures are ensuring that the topic of invertebrate conservation and, most importantly, its relevance to broader conservation issues, is indeed becoming more familiar to more people. The science is now at a crucial phase, with some danger of succumbing to lack of effective global and national co-ordination. It is too important for this to be allowed to occur. Invertebrate conservation is not a 'separate science', despite its general novelty to many people, but part of a more holistic need and concern to conserve the natural world. It is also a need to which individual people can contribute easily on a local scale. As custodians of the greatest proliferation of animal life in all our global ecosystems, invertebrate zoologists must play significant roles in harmonizing the needs of the world's human population with the sustainability of the multitude of other species which share the Earth, and which have dominated its dynamic nature for so long.

FURTHER READING

Gaston, K. J., New, T. R. and Samways, M. J. (ed.) (1994). *Perspectives on insect conservation*. Intercept, Andover.

McNeely, J. A. and Miller, K. R. (ed.) (1984). *National parks, conservation, and development: the role of protected areas in sustaining society*. Smithsonian Press, Washington, DC.

Western, D. and Pearl, M. (ed.). (1989). *Conservation for the twenty-first century*. Oxford University Press, New York.

APPENDIX 1 IUCN RED DATA BOOK STATUS CATEGORIES

1. Extinct. Taxa not definitely located in the wild during the past 50 years.

2. Endangered. Taxa in danger of extinction, and whose survival is unlikely if the causal factors continue operating. This includes taxa that are in immediate danger because of habitat destruction and those that may be already extinct but which have been seen during the past 50 years.

3. Vulnerable. Taxa believed likely to become endangered in the near future if causal factors continue operating.

4. Rare. Taxa with small world populations which are not at present 'endangered' or 'vulnerable', but are at risk.

5. Indeterminate. Taxa known to be in one or other of categories 2–4 but for which there is insufficient information to determine which.

6. Out of danger. Taxa formerly included in one of 2–5 above but which are now considered relatively secure because of effective conservation measures.

The term 'threatened' is a general one to denote taxa in any of categories 2–5.

REFERENCES

Abbott, I. (1990). The influence of fauna on soil structure. In *Animals in primary succession. The role of fauna in reclaimed lands* (ed. J. D. Majer), pp. 39–50. Cambridge University Press, Cambridge.

Ackery, P. R. and Vane-Wright, R. I. (1984). *Milkweed butterflies*. British Museum (Natural History), London/Cornell University Press, New York.

Adamus, P. R. and Clough, G. C. (1978). Evaluating species for protection in natural areas. *Biological Conservation* **13**, 165–78.

Alberch, P. (1993). Museums, collections and biodiversity inventories. *Trends in Ecology and Evolution* **8**, 372–5.

Alcala, A. C., Gomez, E. D., Alcala, L., Cowan, M. E., and Yap, H. T. (1982). Growth of certain corals, molluscs and fish in artificial reefs in the Philippines. *Proceedings of the 4th International Coral Reef Symposium, Manila, Philippines*.

Allan, J. D. and Flecker, A. S. (1993). Biodiversity conservation in running waters. *BioScience* **43**, 32–43.

Allan, P. B. M. (1943). *Talking of moths*. Newtown, Montgomery.

Altaba, C. R. (1990). The last known population of the freshwater mussel *Margaritifera auricularia* (Bivalvia, Unionoida): a conservation priority. *Biological Conservation* **52**, 271–86.

Andersen, A. N. (1990). The use of ant communities to evaluate change in Australian terrestrial ecosystems: a review and a recipe. *Proceedings of the Ecological Society of Australia* **16**, 347–57.

Andersen, A. N. and Yen, A. L. (1985). Immediate effects of fire on ants in the mallee of northwestern Victoria. *Australian Journal of Ecology* **10**, 25–30.

Anderson, E. E. (1989). Economic benefits of habitat restoration: seagrass and the Virginia hard-shell blue crab fishery. *North American Journal of Fisheries Management* **9**, 140–9.

Andersson, H., Coulianos, C.-C., Ehnström, B., Hammarstedt, O., Imby, L., Janzon, L.-Å., Lindelow, Å. and Walden, H. W. (1987). [Threatened invertebrates in Sweden.] *Entomologisk Tidskrift* **108**, 65–75.

André, H. M., Bolly, C., and Lebrun, Ph. (1982). Monitoring and mapping air pollution through an animal indicator: a new and quick method. *Journal of Applied Ecology* **19**, 107–11.

Anon. (1950). Giant snails cross the world. *Wild Life* **12, (2)**, 59–61.

Anon. (1992a). Earthworm's a winner. *The Helix* **30**.

Anon. (1992b). As the worm turns. *The Helix* **27** 6–9.

Anon. (1994). Major new international foundation for conservation to be based in Indonesia. *Conservation Biology* **8**, 15.

Arnold, R. A. (1983). Conservation and management of the endangered Smith's blue butterfly, *Euphilotes enoptes smithi* (Lepidoptera: Lycaenidae). *Journal of Research on the Lepidoptera* **22**, 135–53.

Aston, R. J. (1968). The effect of temperature on the life cycle, growth and fecundity of *Branchiura sowerbyi* (Oligochaeta: Tubificidae). *Journal of Zoology* **154**, 29–40.

Ballard, J. W. O., Olsen, G. J., Faith, D. P., Odgers, W. A., Rowell, D. M. and Atkinson, P. W. (1992). Evidence from 12S ribosomal RNA sequences that onychophorans are modified arthropods. *Science* **258**, 1345–8.

Bardach, J. E., Ryther, J. H., and McLarney, W. O. (1972). *Aquaculture. The farming and husbandry of freshwater and marine organisms*. Wiley-Interscience, New York.

Barnes, R. D. (1989). Diversity of organisms: how much do we know? *American Zoologist* **29**, 1075–84.

Barnes, R. S. K. (1991). Dilemmas in the theory and practice of biological conservation as exemplified by British coastal lagoons. *Biological Conservation* **55**, 315–28.

Barrett, P. (1991). *Rearing wetas in captivity*. Wellington Zoological Gardens, Wellington, New Zealand.

Bauer, G. (1988). Threats to the freshwater pearl mussel *Margaritifera margaritifera* L. in Central Europe. *Biological Conservation* **43**, 239–53.

Bauer, G. (1991). Plasticity in life history traits of the freshwater pearl mussel – consequences for the danger of extinction and for conservation measures. In *Species conservation: a population-biological approach* (ed. A. Seitz, and V. Loeschcke), pp. 103–20. Birkhauser Verlag, Basel.

Baur, A. and Baur, B. (1990). Are roads barriers to dispersal in the land snail *Arianta arbustorum*? *Canadian Journal of Zoology* **68**, 613–17.

Baur, A. and Baur, B. (1992). Effect of corridor width on animal dispersal: a simulation study. *Global Ecology and Biogeography Letters* **2**, 52–6.

Baur, B. and Baur, A. (1993). Climatic warming due to thermal radiation from an urban area as possible cause for the local extinction of a land snail. *Journal of Applied Ecology* **30**, 333–40.

Bayfield, N. (1979). Some effects of trampling on *Molophilus ater* (Meigen) (Diptera, Tipulidae). *Biological Conservation* **16**, 219–32.

Bederman, D. J. (1991). International control of marine 'pollution' by exotic species. *Ecology Law Quarterly* **18**, 677–717.

Beggs, J. R. and Wilson, P. R. (1991). The Kaka *Nestor meridionalis*, a New Zealand parrot endangered by introduced wasps and mammals. *Biological Conservation* **56**, 23–38.

Beirne, B. P. (1955). Natural fluctuations in abundance of British Lepidoptera. *Entomologist's Gazette* **6**, 21–52.

Berman, J., Harris, L., Lambert, W., Buttrick, M., and Dufresne, M. (1992). Recent invasions of the Gulf of Maine: three contrasting ecological histories. *Conservation Biology* **6**, 435–41.

Berry, R. J. (1972). Conservation and the genetical constitution of populations. In *Scientific management of plant and animal communities for conservation* (ed. E. Duffey, and A. S. Watt), pp. 117–206. Blackwell Scientific Publications, Oxford.

Biological Survey of Canada (1994). Terrestrial arthropod biodiversity: planning a study and recommended sampling techniques. *Bulletin of the Entomological Society of Canada* **26**(1), Supplement.

Boatman, N. D., Dover, J. W., Wilson, P. J., Thomas, M. B., and Cowgill, S. E. (1989). Modification of farming practice at field margins to encourage wildlife. In *Biological habitat reconstruction* (ed. G. P. Buckley), pp. 299–311, Belhaven, London.

Bongers, T. (1990). The maturity index: an ecological measure of environmental disturbance based on nematode species composition. *Oecologia* **83**, 14–19.

Bracker, R. G. A. and Bider, J. R. (1982). Changes in terrestrial animal activity of a forest community after an application of aminocarb (Matacil[R]). *Canadian Journal of Zoology* **60**, 1981–97.

Brakefield, P. M. (1991). Genetics and the conservation of invertebrates. In *The scientific management of temperate communities for conservation* (ed. I. F. Spellerberg, F. B. Goldsmith, and M. G. Morris), pp. 45–79. Blackwell Scientific Publications, Oxford.

Bratton, J. H. (ed.) (1991). *British red data books. 3. Invertebrates other than insects*. Joint Nature Conservation Committee, Peterborough.

Briggs, J. C. (1991). Global species diversity. *Journal of Natural History* **25**, 1403–6.

Briggs, J. C. (1994). Species diversity: land and sea compared. *Systematic Biology* **43**, 130–5.

Brinkhurst, R. O. (1993). Future directions in freshwater biomonitoring using benthic macroinvertebrates. In *Freshwater biomonitoring and benthic macroinvertebrates* (ed. D. M. Rosenberg and V. H. Resh), pp. 442–60. Chapman & Hall, New York.

Brinkhurst, R. O. and Cook, D. G. (1974). Aquatic earthworms (Annelida: Oligochaeta). In *Pollution ecology of freshwater invertebrates* (ed. C. W. Hart, and S. L. H. Fuller), pp. 143–56. Academic Press, New York.

Brinkhurst, R. O and Jamieson, B. G. M. (1971). *Aquatic Oligochaeta of the world*. Oliver and Boyd, Edinburgh.

British Dragonfly Society (1992). *Dig a pond for dragonflies*. Purley, Surrey.

Britten, H. B., Brussard, P. F. and Murphy, D. D. (1994). The pending extinction of the Uncompahgre fritillary butterfly. *Conservation Biology* **8**, 86–94.

Brooks, S. J. (1993). Guidelines for invertebrate site surveys. *British Wildlife* (1993), 283–6.

Brower, L. P. and Malcolm, S. B. (1989). Endangered phenomena. *Wings* **14**, 3–10.

Brown, B. (1993). Maine's baitworm fisheries: resources at risk? *American Zoologist* **33**, 568–77.

Brown, I. W. and Fielder, D. R. (ed.) (1991). *The coconut crab: aspects of Birgus latro biology and ecology in Vanuatu*. ACIAR Monograph No. 8. Canberra.

Brown, J. H. (1991). Freshwater prawns. **In** *Production of aquatic animals. crustaceans, molluscs, amphibians and reptiles* (ed. Nash, C. E.), pp. 31–43. Elsevier, Amsterdam.

Brown, K. S. (1991). Conservation of neotropical environments: insects as indicators. **In** *Conservation of insects and their habitats*. (ed. N. M. Collins, and J. A. Thomas), pp. 350–404. Academic Press, London.

Browne, R. A. (1980). Competition experiments between parthenogenetic and sexual strains of the brine shrimp, *Artemia salina*. *Ecology* **61**, 471–4.

Buchsbaum, R. (1966). *Animals without backbones*. Pelican Books, London.

Buchwald, R. (1992). Vegetation and dragonfly fauna – characteristics and examples of biocenological field studies. *Vegetatio* **101**, 99–107.

Buikema, A. L., Niederlehner, B. R., and Cairns, J. (1982). Biological monitoring part IV – toxicity testing. *Water Research* **16**, 239–62.

Burns, G. and Burns, J. (1992). The distribution of Victorian jewel beetles (Coleoptera: Buprestidae). *Occasional Papers from the Museum of Victoria* **5**, 1–53.

Bushnell, J. H. (1974). Bryozoans (Ectoprocta). **In** *Pollution ecology of freshwater invertebrates*. (ed. C. W. Hart, and S. L. H. Fuller), pp. 157–94. Academic Press, New York.

BUTT (Butterflies Under Threat Team). (1986). *The management of chalk grassland for butterflies*. Nature Conservancy Council, Peterborough.

Cairns, J. (1983). Are single species toxicity tests alone adequate for estimating environmental hazard? *Hydrobiologica* **100**, 47–57.

Callaghan, C. A. (1988). Earthworms as ecotoxicological assessment tools. **In** *Earthworms in waste and environmental managment* (ed. Edwards, C. A. and E. F. Neuhauser), pp. 295–301. SPC, The Hague.

Cannon, L. R.G., Goeden, G. B., and Campbell, P.

(1987). Community patterns revealed by trawling in the inter-reef regions of the Great Barrier Reef. *Memoirs of the Queensland Museum* **25**, 45–70.

Carlton, J. T. (1985). Transoceanic and interoceanic dispersal of coastal marine organisms: the biology of ballast water. *Annual Review of Oceanography and Marine Biology* **23**, 313–71.

Carlton, J. T. (1989). Man's role in changing the face of the ocean: biological invasions and implications for conservation of near-shore environments. *Conservation Biology* **3**, 265–73.

Carlton, J. T., Vermeij, G. T., Lindberg, D. R., Carlton, D. A., and Dudley, E. C. (1991). The first historical extinction of a marine invertebrate in an ocean basin: the demise of the Eelgrass Limpet *Lottia alveus*. *Biological Bulletin* **180**, 72–80.

Carpenter, K. E. (1924). A study of the fauna of rivers polluted by lead mining in the Aberystwyth district of Cardiganshire. *Annals of Applied Biology* **11**, 1–23.

Carson, H. L. and Kaneshiro, K. Y. (1976). *Drosophila* of Hawaii: systematics and ecological genetics. *Annual Review of Ecology and Systematics* **7**, 311–46.

Carson, R. (1963). *Silent spring*. Hamish Hamilton, London.

Chandler, J. R. (1970). A biological approach to water quality management. *Journal of Water Pollution Control* **69**, 415–22.

Chelmick, D., Hammond, C., Moore, N., and Stubbs, A. (1980). *The conservation of dragonflies*. Nature Conservancy Council, London.

Chiverton, P. A. and Sotherton, N. W. (1991). The effects on beneficial arthropods of the exclusion of herbicides from cereal crop edges. *Journal of Applied Ecology* **28**, 1027–39.

Clapperton, B. K., Moller, H., and Sandlant, G. R. (1989). Distribution of social wasps (Hymenoptera: Vespidae) in New Zealand in 1987. *New Zealand Journal of Zoology* **16**, 315–23.

Clark, P. B. (1992). Organisation and economics of insect farming. Paper delivered at Invertebrates (Microlivestock) Farming Seminar, La Union, Philippines.

Clark, P. B. and Landford, A. D. (1991). Farming insects in Papua New Guinea. *International Zoo Yearbook* **30**, 127–31.

Clarke, B. and Murray, J. (1969). Ecological genetics and speciation in land snails of the genus *Partula*. *Biological Journal of the Linnean Society* **1**, 31–42.

Clarke, B., Murray, J. and Johnson, M. S. (1984). The extinction of endemic species by a program of biological control. *Pacific Science* **38**, 97–104.

Clarke, D. (1991). Captive-breeding programme for the red-kneed bird-eating spider, *Euathlus smithi*, at London Zoo. *International Zoo Yearbook* **30**, 68–75.

Clarke, G. M. (1992). Fluctuating asymmetry: a technique for measuring developmental stress of genetic and environmental origin. *Annales Zoologicae Fennici* **191**, 31–5.

Clarke, G. M. (1993). Fluctuating asymmetry of invertebrate populations as a biological indicator of environmental quality. *Environmental Pollution* **82**, 207–11.

Clarke, G. M. and McKenzie, L. J. (1992). Fluctuating asymmetry as a quality control indicator for insect mass rearing processes. *Journal of Economic Entomology* **85**, 2045–50.

Clarke, G. M. and Ridsdill-Smith, T. J. (1990). The effect of Avermectin B$_I$ on developmental stability in the bushfly, *Musca vetustissima*, as measured by fluctuating asymmetry. *Entomologia Experimentalis et Applicata* **54**, 265–9.

Clarke, G. M., Brand, G. W., and Whitten, M. J. (1986). Fluctuating asymmetry: a technique for measuring developmental stress caused by inbreeding. *Australian Journal of Biological Sciences* **39**, 145–53.

Coddington, J. A., Griswold, C. E., Davila, D. S., Peñaranda, E., and Larcher, S. F. (1991). Designing and testing sampling protocols to estimate biodiversity in tropical ecosystems. In *The unity of evolutionary biology* (ed. E. C. Dudley), pp. 44–60. Dioscorides Press, Portland, Oregon.

Cody, M. L. (1986). Diversity, rarity, and conservation in Mediterranean climate regions. In *Conservation Biology*. (ed. M. Soulé), pp. 123–52. Sinauer, Sunderland, Mass.

Collier, K. (1993). Review of the status, distribution and conservation of freshwater invertebrates in New Zealand. *New Zealand Journal of Marine and Freshwater Research* **27**, 339–56.

Collins, N. M. (1987a). *Butterfly houses in Britain. The conservation implications*. IUCN, Gland.

Collins, N. M. (1987b). *Legislation to conserve insects in Europe*. Amateur Entomologist's Society, Pamphlet No. **13**, London.

Collins, N. M. (ed.) (1990). *The management and welfare of invertebrates in captivity*. National Federation of Zoological Gardens of Great Britain and Ireland, London.

Collins, N. M. (1991). Insect conservation – priorities for the future. *Antenna* **15**, 73–8.

Collins, N. M. and Morris, M. G. (1985). *Threatened swallowtail butterflies of the world*. IUCN, Gland and Cambridge.

Comps, M., Bonami, J. R., Vago, C., and Campillo, A. (1976). Une vorose de l'huître portugaise (Crassostrea angulata Lmk.). *Comptes Rendues Hebdominales Seances de l'Academie des Sciences, series D, Sciences Naturelles* **282**, 1991–3.

Conant, S. (1988). Saving endangered species by translocation. *BioScience* **38**, 254–7.

CONCOM (Council of Nature Conservation Ministers) (1989). *Australian Statement on Invertebrates*. Canberra.

Cook, D. R. (1974). Water mite genera and subgenera. *Memoirs of the American Entomological Institute* **21**, 1–860.

Cook, D. R. (1986). Water mites from Australia. *Memoirs of the American Entomological Institute* **40**, 1–568.

Cowie, R. H. (1992). Evolution and extinction of Partulidae, endemic Pacific island land snails. *Philosophical Transactions of the Royal Society of London B* **335**, 167–91.

Cram, D. L., Agnebag, J. J., Hampton, I., and Robertson, A. A. (1979). SAS *Protea* cruise 1978. The general results of the acoustics and remote sensing study, with recommendations for estimating the abundance of krill (*Euphausia superba* Dana). *South African Journal of Antarctic Research* **9**, 3–13.

Cranston, P. S. (1990). Biomonitoring and invertebrate taxonomy. *Environmental Monitoring and Assessment* **14**, 265–73.

Cranston, P. S. and Hillman, T. (1992). Rapid assessment of biodiversity using 'Biological Diversity Technicians'. *Australian Biologist* **5**, 144–54.

Crosland, M. J. W. (1991). The spread of the social wasp, *Vespula germanica*, in Australia. *New Zealand Journal of Zoology* **18**, 375–88.

Cullen, J. M. and Delfosse, E. S. (1985). *Echium plantagineum*: catalyst for conflict and change in Australia. In *Proceedings of the VIth International Symposium on Biological Control of Weeds, Vancouver* (ed. E. S. Delfosse), pp. 249–92 Agriculture Canada, Ottawa.

Dance, S. P. (1966). *Shell collecting. An illustrated history*. Faber and Faber, London.

Danks, H. V. (ed.) (1979). Canada and its insect fauna. *Memoirs of the Entomological Society of Canada* **108**.

Dallinger, R., Berger, B., and Birkel, S. (1992). Terrestrial isopods: useful biological indicators of urban metal pollution. *Oecologia* **89**, 32–41.

Davies, M. and Kathirithamby, J. (1986). *Greek insects*. Oxford University Press, New York.

Davies-Colley, R. J., Hickey, C. W., Quinn, J. M., and Ryan, P. A. (1992). Effects of clay discharges on streams. 1. Optical properties and epilithon. *Hydrobiologia* **248**, 215–34.

Davis, B. N. K., Lakhani, K. H., and Yates, T. J. (1991). The hazards of insecticides to butterflies of field margins. *Agriculture, Ecosystems, Environment* **36**, 151–61.

Davis, D. L. (1985). *Hemiphlebia mirabilis* Selys: some notes on distribution and conservation status (Zygoptera: Hemiphlebiidae). *Odonatologica* **14**, 331–9.

Dayton, P. K. (1972). Towards an understanding of community resilience and the potential effects of enrichment to the benthos of McMurdo Sound, Antarctica. In *Proceedings of the Colloquium on Conservation Problems in Antarctica*, (ed. B. C. Parker), pp. 81–96. Allen Press, Lawrence, Kansas.

De Foliart, G. R. (1989). The human use of insects as food. *Bulletin of the Entomological Society of America* **35**, 22–35.

Den Boer, P. J. (1968). Spreading of risk and the stabilisation of animal numbers. *Acta Biotheoretica* **18**, 165–94.

Diamond, J. M. (1975). The island dilemma: lessons of modern biogeographic studies for the design of natural reserves. *Biological Conservation* **7**, 129–46.

Di Castri, F. (1990). On invading species and invaded ecosystems: the interplay of historical chance and biological necessity. In *Biological invasions in Europe and the Mediterranean Basin* (ed. F. di Castri, A. J. Hansen, and M. Debussche), pp. 3–16. Kluwer, Dordrecht.

Di Castri, F., Vernhes, J. R., and Younés, T. (1992). Inventorying and monitoring biodiversity. A proposal for an international network. *Biology International*, Special Issue **27**, 1–28.

Disney, R. H. L. (1975). *Environment and creation*. Chester House, London.

Disney, R. H. L. (1982). Rank methodists. *Antenna* **6**, 198.

Disney, R. H. L. (1986*a*). Assessments using invertebrates: posing the problem. In *Wildlife conservation evaluation*. (ed. M. B. Usher), pp. 271–93. Chapman & Hall, London.

Disney, R. H. L. (1986*b*). Inventory surveys of insect faunas: discussion of a particular attempt. *Antenna* **10**, 112–16.

Ditlhogo, M. K. M., James, R., Laurence, B. R., and Sutherland, W. J. (1992). The effects of conservation management on reed beds. I. The invertebrates. *Journal of Applied Ecology* **29**, 265–76.

Dix, T. G. (1990). The Pacific Oyster, *Crassostrea gigas*, in Australia. In *Estuarine and marine bivalve mollusk culture* (ed. W. Menzel), pp. 315–18. CRC Press, Boca Raton, Florida.

Donald, D. B. (1980). Deformities in Capniidae (Plecoptera) from Bow River, Alberta. *Canadian Journal of Zoology* **58**, 682–6.

Donnelly, D. and Giliomee, J. (1985). Community structure of epigaeic ants in a pine plantation and in newly burnt fynbos. *Journal of the Entomological Society of southern Africa* **48**, 259–65.

Dover, J. W. (1989). The use of flowers by butterflies foraging in cereal field margins. *Entomologist's Gazette* **40**, 283–91.

Dover, J. W. (1990). Butterflies and wildlife corridors. *Game Conservation Review* (1988) **20**, 54–6.

Dover, J. W. (1991). The conservation of insects on arable farmland. In *The conservation of insects and their habitats* (ed. N. M. Collins and J. A. Thomas), pp. 294–318. Academic Press, London.

Dover, J., Sotherton, N., and Gobbett, K. (1990). Reduced pesticide inputs on cereal field margins: the effects on butterfly abundance. *Ecological Entomology* **15**, 17–24.

Drake, J. A., *et al.* (ed.) (1989). *Biological invasions: a global perspective*. Wiley, New York.

Duffey, E. (1975). The effects of human trampling on the fauna of grassland litter. *Biological Conservation* **7**, 255–74.

Duffey, E. (1977). The re-establishment of the large copper butterfly *Lycaena dispar batava* Obth. on Woodwalton Fen National Nature Reserve, Cambridgeshire, England, 1969–1973. *Biological Conservation* **12**, 143–58.

Duffey, E. (1993). The Large Copper, *Lycaena dispar*. In *Conservation biology of the Lycaenidae* (ed. T. R. New), pp. 81–2. IUCN, Gland.

Eberhard, S. M., Richardson, A. M. M., and Swain, R. (1991). *The invertebrate cave fauna of Tasmania*. Zoology Department, University of Tasmania, Hobart.

Ehrenfeld, D. W. (1970). *Biological conservation*. Holt, Rinehard and Winston, New York.

Ehrlich, P. R. and Murphy, D. D. (1987). Conservation lessons from long-term studies of checkerspot butterflies. *Conservation Biology* **1**, 122–31.

Elliott, W. R. (1991). Cave fauna conservation in Texas. In *Proceedings of the 1991 National Cave Management Symposium* (ed. D. G. Foster), pp. 1–14. American Cave Conservation Association, Horse Cave, Kentucky.

Elmes, G. W. and Thomas, J. A. (1992). Complexity of species conservation in managed habitats: interaction between *Maculinea* butterflies and their ant hosts. *Biodiversity and Conservation* **1**, 155–69.

Elton, C. S. (1958). *The ecology of invasions by animals and plants*. Methuen, London.

Endersby, I. D. (1993). A new locality for *Hemiphlebia mirabilis* Selys (Odonata: Hemiphlebiidae). *Victorian Entomologist* **23**, 4–5.

ESV (Entomological Society of Victoria) (1986). *Preliminary distribution maps of butterflies in Victoria*. Entomological Society of Victoria, Melbourne.

Erhardt, A. and Thomas, J. A. (1991). Lepidoptera as indicators of change in the semi-natural grasslands of lowland and upland Europe. In *The conservation of insects and their habitats* (ed. N. M. Collins, and J. A. Thomas), pp. 213–36. Academic Press, London.

Erwin, T. L. (1982). Tropical forests: their richness in Coleoptera and other arthropod species. *Coleopterists Bulletin* **36**, 74–5.

Erwin, T. L. (1991). How many species are there?: revisited. *Conservation Biology* **5**, 330–3.

European Community (1986). *European Charter for Invertebrates*. Council for Europe, Strasbourg.

European Community (1992). Council Directive 92/43/EEC on the conservation of natural habitats and of wild fauna and flora. *Official Journal of the European Communities* **L206**, 7–50.

Everson, I. (1981). Antarctic krill. In *Biological investigations of marine Antarctic systems and stocks. II. Selected contributions* (ed. S. Z. El-Sayed), pp. 31–46. Scott Polar Research Institute, Cambridge.

Eyre, M. D. and Luff, M. L. (1989). A preliminary classification of European grassland habitats using carabid beetles. In *The role of ground beetles in ecological and environmental studies* (ed. N. E. Stork), pp. 227–36. Intercept, Andover, Hants.

Eyre, M. D. and Rushton, S. P. (1989). Quantification of conservation criteria using invertebrates. *Journal of Applied Ecology* **26**, 159–71.

Eyre, M. D., Ball, S. G., and Foster, G. N., (1986*a*). An initial classification of the habitats of aquatic Coleoptera in north-east England. *Journal of Applied Ecology* **23**, 841–52.

Eyre, M. D., Rushton, S. P., Luff, M. L., Ball, S. G., Foster, G. N. and Topping, C. J. (1986*b*). *The use of invertebrate community data in environmental assessment*. University of Newcastle-upon-Tyne, Agricultural Environmental Research Group.

Ferrar, A. A. (1989). The role of Red Data Books in conserving biodiversity. In *Biotic diversity in southern Africa* (ed. B. J. Huntley), pp.136–47. Oxford University Press, Cape Town.

Fiedler, P. L., Leidy, R. A., Laven, R. D., Gershenz, N., and Saul, L. (1993). The contemporary paradigm in ecology and its implications for endangered species conservation. *Endangered Species Update* **10**, (314): 7–12.

Fitt, W. K. (1990). Mariculture of giant clams. In *Estuarine and marine bivalve mollusk culture* (ed. W. Menzel), pp. 283–295. CRC Press, Boca Raton, Florida.

Fitter, R. and Fitter, M. (1987). *The road to extinction*. IUCN, Gland and Cambridge.

Fletcher, A. R. (1979). Effects of *Salmo trutta* on *Galaxias olidus* and macroinvertebrates in stream communities. M.Sc. Thesis, Monash University, Melbourne.

Fletcher, A. R. (1986). Effects of introduced fish in Australia. In *Limnology in Australia* (ed. P. De Deckker, and W. D. Williams), pp. 231–8. CSIRO, Melbourne, W. Junk, Dordrecht.

Foltz, D. W., Ochman, H., and Selander, R. K. (1984). Genetic diversity and breeding systems in terrestrial slugs of the families Limacidae and Arionidae. *Malacologia* **25**, 593–605.

Ford, B. C., *et al.* (1988). Cesium–134 and cesium–137 in honey bees and cheese samples collected in the US after the Chernobyl accident. *Chemosphere* **17**, 1153.

Ford, E. B. (1975). *Ecological genetics* (4th edn). Chapman & Hall, London.

Ford, M. J. (1982). *The changing climate. Responses of the natural flora and fauna*. George Allen & Unwin, London.

Forman, R. T. T. and Godron, M. (1986). *Landscape ecology*. Wiley, New York.

Foster, G. N. (1991). Conserving insects of aquatic and wetland habitats, with special reference to beetles. In *The conservation of insects and their habitats*. (ed. N. M. Collins, and J. A. Thomas), pp. 237–62. Academic Press, London.

France, R. L. and Collins, N. C. (1993). Extirpation of crayfish in a lake affected by long-range anthropogenic acidification. *Conservation Biology* **7**, 184–8.

Frankel, O. (1976). The biological structure of the landscape. In *Man and Landscape in Australia. Towards an ecological vision* (ed. G. Seddon, and M. Davis), pp. 49–62. Australian Government Publishing Service, Canberra.

Frankel, O. H. and Soulé, M.E (1981). *Conservation and evolution*. Cambridge University Press, Cambridge.

Freitag, R. (1979). Carabid beetles and pollution. In *Carabid beetles: their evolution, natural history and classification* (ed. T. L. Erwin, G. E. Ball, and D. R. Whitehead), pp. 507–21. W. Junk, The Hague.

Fry, G. L. A. (1991). Conservation in agricultural ecosystems. In *The scientific management of temperate communities for conservation (ed. I. F. Spellerberg, F. B. Goldsmith, and M. G. Morris), pp. 415–43. Blackwell Scientific Publications, Oxford.

Fry, R. and Lonsdale, D. (ed.) (1991). *Habitat conservation for insects – a neglected green issue*. The Amateur Entomologists' Society, Middlesex, UK.

Fuller, S. L. H. (1974). Clams and mussels (Mollusca: Bivalvia). In *Pollution ecology of freshwater invertebrates*. (ed. C. W. Hart, Jr., and S. L. H. Fuller), pp. 215–73. Academic Press, New York.

Gagné, W. C. and Howarth, F. G. (1984). Conservation status of endemic Hawaiian Lepidoptera. *Proceedings of the 3rd Congress of European Lepidopterology, Cambridge*, **1982**; pp. 74–84.

Gall, L. F. (1984). Population structure and recommendations for conservation of the narrowly endemic alpine butterfly, *Boloria acrocnema* (Lepidoptera: Nymphalidae). *Biological Conservation* **28**, 111–38.

Gambino, P., Medeiros, A. C., and Loope, L. L. (1990). Invasion and colonisation of upper elevations on east Maui (Hawaii) by *Vespula pensylvanica* (Hymenoptera: Vespidae). *Annals of the Entomological Society of America* **83**, 1088–95.

Gaston, K. J. (1991). Estimates of the near-imponderable: a reply to Erwin. *Conservation Biology* **5**, 564–6.

Gauch, H. G. (1982). *Multivariate analysis in community ecology*. Cambridge University Press, Cambridge.

Geddes, M. C. and Williams, W. D. (1987). Comments on *Artemia* introductions and the need for conservation. In Artemia *research and its applications. 3. Ecology, culturing, use in aquaculture* (ed. P., Sorgeloos, D. A. Bengston, W. Decleir, and E. Jaspers), pp. 19–26. Universa Press, Welteren.

Ghiselin, M. T. (1984). *Peripatus* as a living fossil. In *Living fossils* (ed. N. Eldredge, and S. M. Stanley), pp. 214–7. Springer, New York.

Gillespie, R. C. and Reimer, N. (1993). The effect of alien predatory ants (Hymenoptera: Formicidae) on Hawaiian endemic spiders (Araneae: Tetragnathidae). *Pacific Science* **47**, 21–33.

Glynn, P. W. (1993). Coral reef bleaching: ecological perspectives. *Coral Reefs* **12**, 1–17.

Goats, G. C. and Edwards, C. A. (1988). The prediction of field toxicity of chemicals to earthworms by laboratory methods. In *Earthworms in waste and environmental management* (ed. C. A. Edwards, and E. F. Neuhauser), pp. 283–94. SPB, The Hague.

Grassle, J. F. (1991). Deep sea benthic biodiversity. *BioScience* **41**, 464–9.

Grassle, J. F. and Maciolek, N. J. (1992). Deep-sea species richness: regional and local diversity estimates from quantitative bottom samples. *American Naturalist* **139**, 313–41.

Gray, C. A., McDonall, V. C., and Reid, D. D. (1990). By-catch from prawn trawling in the Hawkesbury River, New South Wales: species composition, distribution and abundance. *Australian Journal of Marine and Freshwater Research* **41**, 13–26.

Greatorex-Davis, J. N., Sparks, T. H., Hall, M. L., and Marrs, R. H. (1993). The influence of shade on butterflies in rides of conifered lowland woods in southern England and implications for conservation management. *Biological Conservation* 63, 31–41.

Green, P. C. (1989). The use of Trichoptera as indicators of conservation value: Hertfordshire gravel pits. *Journal of Environmental Management* 29, 95–104.

Greenslade, P. J. M. (1978). Ants. In *The physical and biological features of Kunnoth Paddock in Central Australia* (ed. W. A. Low). *Technical Paper* No. 4, CSIRO Division of Land Resources Management.

Greenslade, P. J. M. (1983). Adversity selection and the habitat templet. *American Naturalist* 122, 352–65.

Groombridge, B. (ed.) (1992) *Global biodiversity; status of the Earth's living resources*. World Conservation Monitoring Centre, Cambridge/ Chapman & Hall, London.

Hadfield, M. G. (1986). Extinction in Hawaiian achatinelline snails. *Malacologia* 27, 67–81.

Hadfield, M. G. and Mountain, B. S. (1981). A field study of a vanishing species, *Achatinella mustelina* (Gastropoda, Pulmonata), in Waianae Mountains of Oahu. *Pacific Science* 34, 345–58.

Hall, R. J., Likens, G. E., Fiance, S. B., and Hendrey, G. R. (1980). Experimental acidification of a stream in the Hubbard Brook Experimental Forest, New Hampshire. *Ecology* 61, 976–89.

Hammond, C. (1983). The courtly crickets. *Arts of Asia* (March–April), 81–7.

Hancocks, D. (1992). Why zoos have forgotten the little animals that run the world. In Sonoran Arthropod Studies, Inc. 1991 Annual Report.

Harman, W. N. (1974). Snails (Mollusca: Gastropoda). In *Pollution ecology of freshwater invertebrates* (ed. C. W. Hart, and S. L. H. Fuller), pp. 275–312. Academic Press, New York.

Harriott, V. J. and Fisk, D. A. (1988). Recruitment patterns of scleractinian corals: a study of three reefs. *Australian Journal of Marine and Freshwater Research* 39, 409–16.

Harris, A. N. and Poiner, I. R. (1990). Bycatch of the prawn fishery of Torres Strait; composition and partitioning of the discards into components that float or sink. In *The effects of fishing* (ed. W. Craik, J. Glaister, and I. Poiner), pp. 37–52. CSIRO, Melbourne.

Harris, P. (1973). The selection of effective agents for the biological control of weeds. *Canadian Entomologist* 105, 1495–1503.

Harris, R. J. (1991). Diet of the wasps *Vespula vulgaris* and *V. germanica* in honeydew beech forest of the South Island, New Zealand. *New Zealand Journal of Zoology* 18, 159–69.

Harrison, F. W. (1974). Sponges (Porifera: Spongillidae). In *Pollution ecology of freshwater invertebrates* (ed. C. W. Hart, and S. L. H. Fuller), pp. 29–66. Academic Press, New York and London.

Hart, C. W. and Fuller, S. L. H. (1974). *Pollution ecology of freshwater invertebrates*. Academic Press, New York.

Harvey, M. S. and Yen, A. L. (1989). *Worms to wasps. An illustrated guide to Australia's terrestrial invertebrates*. Oxford University Press, Melbourne.

Haslett, J. R. (1988). Assessing the quality of alpine habitats: hoverflies (Diptera: Syrphidae) as bio-indicators of skiing pressure on alpine meadows in Austria. *Zoologisches Anzeiger* 220, 179–84.

Hatekeyama, S., Sugaya, Y., Satake, K., Miyashita, M., and Fukushima, S. (1991). Microinvertebrate communities in heavy metal-polluted rivers in the Shikoku district of Japan. *Verhandlungen der Internationalen Vereinigung fur Theoretische und Angewandte Limnologie.* 24, 2220–7.

Hawksworth, D. L. and Mound, L. A. (1991). Biodiversity databases: the crucial significance of collections. In *The biodiversity of microorganisms and invertebrates: its role in sustainable agriculture* (ed. D. L. Hawksworth), pp. 17–29. CAB International, Wallingford.

Heath, J. (1981). *Threatened Rhopalocera (butterflies) in Europe*. Council of Europe. Nature and Environment Series No. 23. Strasbourg.

Heath, J. and Leclerq, J. (ed.) (1981). *Provisional atlas of the invertebrates of Europe*. Institute of Terrestrial Ecology, Cambridge and Faculty des Sciences Agronomiques, Gembloux.

Heath, J., Pollard, E., and Thomas, J. A. (1984). *Atlas of butterflies in Britain and Ireland*. Viking, Harmondsworth.

Hebert, P. D. N., Billington, N., Finston, T. L., Boileau, M. G., Beaton, M. J., and Barrette, R. J. (1991). Genetic variation in the onychophoran *Plicatoperipatus jamaicensis*. *Heredity* 67, 221–30.

Heliövaara, K. and Väisänen, R. (1993). *Insects and Pollution*. CRC Press, Boca Raton.

Hellawell, J. M. (1986). *Biological indicators of freshwater pollution and environmental management*. Elsevier, London

Henning, S. F. and Henning, G. A. (1989). *South African red data book - butterflies*. Council for Scientific and Industrial Research, Pretoria.

Héral, M. and Deslous-Paoli, J. M. (1990). Oyster culture in European countries. In *Estuarine and marine bivalve mollusk culture* (ed. W. Menzel), pp. 153–190. CRC Press, Boca Raton, Florida.

Hill, B. J. and Wassenberg, T. J. (1990). Fate of discards from prawn trawlers in Torres Strait. In *The effects of fishing* (ed. W. Craik, J. Glaister, and I. Poiner), pp. 53–64. CSIRO, Melbourne.

Hockey, P. A. R. and Bosman, A. L. (1986). Man as an intertidal predator in Transkei: disturbance, community convergence and management of a natural food resource. *Oikos* **46**, 3–14.

Holdgate, M. W. (1991). Conservation in a world context. In *The scientific management of temperate communities for conservation* (ed. I. F. Spellerberg, F. B. Goldsmith, and M. G. Morris), pp. 1–26. Blackwell Scientific Publications, Oxford.

Holditch, D. M. (1987). The dangers of introducing alien animals with particular reference to crayfish. In *Freshwater crayfish. 7* (ed. P. Goeldlin de Tiefenau), pp. 15–30. International Symposium of Astacology, Lausanne, Switzerland.

Hölldobler, B. and Wilson, E. O. (1990). *The Ants*. Harvard University Press, Cambridge, Mass.

Holloway, J. D. (1976). *Moths of Borneo with special reference to Mount Kinabalu*. Malayan Nature Society, Kuala Lumpur.

Holloway, J. D. and Stork, N. E. (1991). The dimensions of biodiversity: the use of invertebrates as indicators of human impact. In *The biodiversity of microorganisms and invertebrates: its role in sustainable agriculture* (ed. D. L. Hawksworth), pp. 37–61. CAB International, Wallingford.

Hooper, M. D. (1971). The size and surroundings of nature reserves. In *The scientific management of animal and plant species for conservation* (ed. E. Duffey, and A. S. Watt), pp. 555–61. Blackwell Scientific Publications, Oxford.

Hopkin, S. P., Hardisty, G. N., and Martin, M. H. (1986). The woodlouse *Porcellio scaber* as a biological indicator of zinc, cadmium, lead and copper pollution. *Environmental Pollution* **11**(B), 271–90.

Hopkin, S. P., Hames, C. A. C., and Bragg, S. (1989). Terrestrial isopods as biological indicators of zinc pollution in the Reading area, south east England. *Monitori Zoologici Italia* **4**, 477–88.

Hopkins, P. R. and Webb, N. R. (1984). The composition of the beetle and spider faunas on fragmented heathlands. *Journal of Applied Ecology* **21**, 935–46.

Hopper, D. R. and Smith, B. D. (1992). Status of tree snails (Gastropoda: Partulidae) on Guam, with a resurvey of sites studied by H. E. Crampton in 1920. *Pacific Science* **46**, 77–85.

Horwitz, P. (1990*a*). The translocation of freshwater crayfish in Australia: potential impact, the need for control and global relevance. *Biological Conservation* **54**, 291–305.

Horwitz, P. (1990*b*). *The conservation status of Australian freshwater Crustacea*. Report Series No. 14, Australian National Parks and Wildlife Service, Canberra.

Horwitz, P. and Hamr, P. (1988). An assessment of the Caroline Creek Freshwater Crayfish Reserve in northern Tasmania. *Papers and Proceedings of the Royal Society of Tasmania* **122**, 69–72.

Howarth, F. G. (1979). The no-eyed big-eyed wolf spider (*Adelocosa anops* Gertsch). Proposal to US Fish and Wildlife Service.

Howarth, F. G. (1981). The conservation of cave invertebrates. *Proceedings of the 1st International Cave Management Symposium*, pp. 57–64.

Howarth, F. G. (1983). Classical biocontrol: panacea or Pandora's box? *Proceedings of the Hawaiian Entomological Society* **24**, 239–44.

Howarth, F. G. (1986). Impact of alien land arthropods and mollusks on native plants and animals in Hawaii. In *Hawai'i's terrestrial ecosystems: preservation and management* (ed. C. P. Stone, and J. M. Scott), pp. 149–79. University of Hawaii, Honolulu.

Howarth, F. G. (1991). Environmental impacts of classical biological control. *Annual Review of Entomology* **36**, 485–509.

Howarth, F. G. and Ramsay, G. W. (1991). The conservation of island insects and their habitats. In *The conservation of insects and their habitats* (ed. N. M. Collins, and J. A. Thomas), pp. 71–107. Academic Press, London.

Hoy, M. A. (1992). Biological control of arthropods:

genetic engineering and environmental risks. *Biological Control* **2**, 166–70.

Huhta, V. (1979). Evaluation of different similarity indices as measures of succession in arthropod communities of the forest floor after clear-cutting. *Oecologia* **41**, 11–23.

Huner, J. V. (1991). Aquaculture of freshwater crayfish. In *Production of aquatic animals. Crustaceans, molluscs, amphibians and reptiles* (ed. C. E. Nash), pp. 45–66. Elsevier, Amsterdam.

Hutchings, P. A. (1990). Review of the effects of trawling on macrobenthic epifaunal communities. In *The effects of fishing* (ed. W. Craik, J. Glaister, and I. Poiner), pp. 111–20. CSIRO, Melbourne.

Hutchings, P. A., van der Velde, J. T., and Keable, S. J. (1987). *Guidelines for the conduct of surveys for detecting introductions of non-indigenous marine species by ballast water and other vectors – and a review of marine introductions to Australia*. Occasional Reports of the Australian Museum No. 3, Sydney.

Hutchinson, G. E. (1957). Concluding remarks. *Cold Spring Harbour Symposium on Quantitative Biology* **22**, 415–27.

Hutson, B. R. (1990). The role of fauna in nutrient turnover. In *Animals in primary succession. The role of fauna in reclaimed lands* (ed. J. D. Majer), pp. 51–70. Cambridge University Press, Cambridge.

Hynes, H. B. N. (1960). *The biology of polluted waters*. Liverpool University Press, Liverpool.

Hynes, H. B. N. (1961). The effect of sheep-dip containing the insecticide BHC on the fauna of a small stream, including *Simulium* and its predators. *Annals of Tropical Medicine and Parasitology* **55**, 192–6.

Hynes, H. B. N. and Roberts, F. W. (1962). The biological effects of synthetic detergents in the River Lee, Hertfordshire. *Annals of Applied Biology* **50**, 779–90.

Hynes, H. B. N. and Williams, T. R. (1962). The effect of DDT on the fauna of a central African stream. *Annals of Tropical Medicine and Parasitology* **56**, 78–91.

IUCN (International Union for the Conservation of Nature and Natural Resources) (1990*a*). *Red list of threatened animals*. IUCN, Gland and Cambridge.

IUCN (International Union for the Conservation of Nature and Natural Resources) (1990*b*).

IUCN directory of South Asian protected areas. IUCN, Gland and Cambridge.

IUCN, UNEP, WWF (International Union for the Conservation of Nature and Natural Resources, United Nations Environmental Programme, World Wildlife Fund for Nature) (1980). *World conservation strategy*. IUCN, Gland.

IUCN, UNEP, WWF (International Union for the Conservation of Nature and Natural Resources, United Nations Environmental Programme, World Wildlife Fund for Nature) (1991). *Caring for the world*. IUCN, Gland.

Jackson, J. B. C. (1991). Adaptation and diversity of reef corals. *BioScience* **41**, 475–82.

Jamieson, G. S. (1993). Marine invertebrate conservation: evaluation of fisheries over-exploitation concerns. *American Zoologist* **33**, 551–67.

Janzen, D. H. (1986). The future of tropical biology. *Annual Review of Ecology and Systematics* **17**, 305–23.

Janzen, D. H. (1987). Insect diversity of a Costa Rican dry forest: why keep it, and how? *Biological Journal of the Linnean Society* **30**, 343–56.

JCCBI (Joint Committee for the Conservation of British Insects) (1986). Insect re-establishment – a code of conservation practice. *Antenna* **10**, 13–18.

Jenkins, D. W. (1971). Global biological monitoring. In *Man's impact on terrestrial and oceanic ecosystems* (ed. W. H. Matthews), pp. 351–70. Massachussets Institute of Technology Press. Cambridge, Mass.

Johannes, R. E. (1978). Traditional marine conservation methods in Oceania and their demise. *Annual Review of Ecology and Systematics* **9**, 349–64.

Johnson, R. K., Wiederholm, T., and Rosenberg, D. M. (1993). Freshwater biomonitoring using individual organisms, populations, and species assemblages of benthic macroinvertebrates. In *Freshwater biomonitoring and benthic macroinvertebrates* (ed. D. M. Rosenberg, and V. H. Resh), pp. 40–158. Chapman & Hall, New York.

Jones, J. R. E. (1938). The relative sensitivity of aquatic species to lead in solution. *Journal of Animal Ecology* **7**, 287–9.

Jones, R. C. and Clark, C. C. (1987). Impact of watershed urbanisation on stream insect communities. *Water Resources Bulletin* **23**, 1047–55.

Kawakatsu, M. and Itô, T. (1963). [Report on the

ecological survey of freshwater planarians in the Ishizuchi mountain range, Shikoku]. *Japanese Journal of Ecology* **13**, 231–4 (in Japanese).

Kenk, R. (1974). Flatworms (Platyhelminthes: Tricladida). In *Pollution ecology of freshwater invertebrates* (ed. C. W. Hart, and S. L. H. Fuller), pp. 67–80. Academic Press, New York.

Keough, M. J., Quinn, G. P., and King, A. (1993). Correlations between human collecting and intertidal mollusc populations on rocky shores. *Conservation Biology* **7**, 378–90.

Kerney, M. and Stubbs, A. (1980). *The conservation of snails, slugs, and freshwater mussels*. Nature Conservancy Council, Shrewsbury.

Key, K. H. L. (1978). *The conservation status of Australia's insect fauna*. Occasional Paper No. 1, Australian National Parks and Wildlife Service, Canberra.

Key, K. H. L. (1982). Species, parapatry, and the morabine grasshoppers. *Systematic Zoology* **30**, 425–58.

Kimura, T., Keegan, H. L., and Haberkorn, T. (1967). Dehydrochlorination of DDT by Asian blood-sucking leeches. *American Journal of Tropical Medicine and Hygiene* **16**, 688–90.

Kinzie, R. A. (1992). Predation by the introduced carnivorous snail *Euglandina rosea* (Ferussac) on endemic aquatic lymnaeid snails in Hawaii. *Biological Conservation* **60**, 149–55.

Kirby, P. (1992). *Habitat management for invertebrates: a practical handbook*. Royal Society for the protection of Birds, Sandy, Bedfordshire.

Köhler, H.-R., Storch, V., and Alberti, G. (1992). The impact of lead of the assimilation efficiency on laboratory-held Diplopoda (Arthropoda) preconditioned in different environmental situations. *Oecologia* **90**, 113–19.

Kondo, Y. (1980). Endangered land snails. Pacific. [Paper delivered to IUCN unpublished.]

Knox, G. A. (1984). The key role of krill in the ecosystem of the southern ocean with special reference to the Convention on the Conservation of Antarctic Marine Living Resources. *Ocean Management* **9**, 113–56.

Kremen, C. (1992). Assessing the indicator properties of species assemblages for natural areas monitoring. *Ecological Applications* **2**, 203–17.

Kremen, C., Colwell, R. K., Erwin, T. L., Murphy, D. D., Noss, R. F., and Sanjayan, M. A.(1993). Terrestrial arthropod assemblages; their use in conservation planning. *Conservation Biology* **7**, 796–808.

Kreutzweiser, D. P. and Sibley, P. K. (1991). Invertebrate drift in a headwater stream treated with Permethrin. *Archives of Environmental Contamination and Toxicology* **20**, 330–6.

Kudrna, O. (1986). *Butterflies of Europe. 8. Aspects of the conservation of butterflies in Europe*. Aula-Verlag, Weisbaden.

Lande, R. and Barrowclough, G. F. (1987). Effective population size, genetic variation, and their use in population management. In *Viable populations for conservation* (ed. M. Soulé), pp. 87–123. Cambridge University Press, Cambridge.

LaRoe, E. T. (1993). Implementation of an ecosystem approach to endangered species conservation. *Endangered Species Update* **10**, 3–6.

Lasebikan, B. A. (1975). The effect of clearing on the soil arthropods of a Nigerian rain forest. *Biotropica* **7**, 84–9.

Lattin, J. D. (1993). Arthropod diversity and conservation in old-growth northwestforests. *American Zoologist* **33**, 578–87.

Leader-Williams, N., Harrison, J., and Green, M. J. B. (1990). Designing protected areas to conserve natural resources. *Science Progress* **74**, 189–204.

Lee, W. L., Bell, B. M., and Sutton, J. F. (1982). *Guidelines for acquisition and management of biological specimens*. Association of Systematics Collections, Lawrence, Kansas.

Lees, D. (1989). Practical considerations and techniques in the captive breeding of insects for conservation purposes. *Entomologist* **108**, 77–96.

Lehmkuhl, D. M., Danks, H. V. Behan-Pelletier, V. M., Larson, D. J., Rosenberg, D. M., and Smith, I. M. (1984). Recommendations for the appraisal of environmental disturbance: some general guidelines and the value and feasibility of insect studies. *Bulletin of the Entomological Society of Canada* **16**, (3), (Suppl.)

Lesica, P. and Allendorf, F. W. (1992). Are small populations of plants worth preserving? *Conservation Biology* **6**, 135–9.

Lewis, T. (1969). The diversity of the insect fauna in a hedgerow and neighbouring fields. *Journal of Applied Ecology* **6**, 453–8.

Liddle, M. J. (1991). Recreation ecology: effects of trampling on plants and corals. *TREE* **6**, 13–17.

Lindqvist, O. V. (1987). Restoration of native European stocks. In *Freshwater Crayfish* 7 (ed.

P. Goeldlin de Tiefenau), pp. 6–12. International Symposium of Astacology, Lausanne, Switzerland.

Ljüngstrom, P. O. (1969). On the earthworm genus *Udeina* in South Africa. *Zoologisches Anzeiger* **182**, 370–9.

Ljüngstrom, P. O. (1972). Taxonomical and ecological note on the earthworm genus *Udeina* and a requiem for the South African acanthodrilines. *Pedobiologia* **12**, 100–10.

Lockwood, J. A. (1987). The moral standing of insects and the ethics of extinction. *Florida Entomologist* **70**, 70–89.

Lockwood, J. A. and De Brey, L. D. (1990). A solution for the sudden and unexplained extinction of the Rocky Mountain grasshopper (Orthoptera: Acrididae). *Environmental Entomology* **19**, 1194–1205.

Lofgren, C. S. (1986). The economic importance and control of imported fire ants in the United States. In *Economic impact and control of social insects* (ed. S. B. Vinson), pp. 227–56. Praeger, New York.

Lofs-Holmin, A. and Bostrom, U. (1988). The use of earthworms and other soil animals in pesticide testing. In *Earthworms in waste and environmental management* (ed. C. A. Edwards, and E. F. Neuhausers), pp. 303–13. SPB, The Hague.

Lower, W. R. and Kendall, R. J. (1990). Sentinel species and sentinel bioassay. In *Biomarkers of environmental contamination* (ed. J. F. McCarthy, and L. R. Shugart), pp. 309–31. Lewis Publishers.

Lowman, F. G. (1979). US Mussel Watch Program. The mussel watch biological monitoring research program. In *Monitoring environmental materials and specimen banking* (ed. N.-P. Luepke), pp. 3982–6. Martinus Nijhoff, The Hague.

Lubchenko, J. *et al.* (1991). The sustainable biosphere initiative: an ecological research agenda. *Ecology* **72**, 371–412.

Luepke, N.-P. (ed.) (1979) *Monitoring environmental materials and specimen banking*. Martinus Nijhoff, The Hague.

Luff, M. L. (ed.) (1987). *The use of invertebrates in site assessment for conservation*. Agricultural Research Group, University of Newcastle-upon-Tyne.

Macan, T. T. (1974). Freshwater invertebrates. In *The changing flora and fauna of Britain* (ed. D. L. Hawksworth), pp. 143–55. Academic Press, London.

McArdle, B. H. (1990). When are rare species not there? *Oikos* **57**, 276–7.

MacArthur, R. H. and Wilson, E. O. (1967). *The theory of island biogeography*. Princeton University Press, Princeton, NJ.

McCahon, C. P. and Pascoe, D. (1990). Episotic pollution: causes, toxicological effects and ecological significance. *Functional Ecology* **4**, 375–83.

Mace, G. A. and Lande, R. (1991). Assessing extinction threats: towards a reevaluation of IUCN threatened species categories. *Conservation Biology* **5**, 148–57.

MacKay, W. P., Artemio Rebeles, M., Arredondo B., H. C., Rodriguez R., A. D., Gonzalez, D. A. and Vinson, S. B. (1991). Impact of the slashing and burning of a tropical rain forest on the native ant fauna (Hymenoptera: Formicidae). *Sociobiology* **18**, 257–68.

McNeely, J. A. and Miller, K. R. (ed.) (1984). *National parks, conservation and development. The role of protected areas in sustaining society*. Smithsonian Institution Press, Washington.

McNeely, J. A., Miller, K. R., Reid, W. V., Mittermeier, R. A., and Werner, T. B. (1989). *Conserving the world's biological diversity*. IUCN, Gland.

Mader, H.-J. (1984). Animal habitat isolation by roads and agricultural fields. *Biological Conservation* **29**, 81–96.

Mader, H.-J. (1986). The succession of carabid species in a brown coal mining area and the influence of afforestation. In *Carabid beetles* (ed. P. J. den Boer, M. L. Luff, D. Mossakowski, and F. Weber), pp. 497–508. Gustav Fischer, Stuttgart.

Mader, H.-J., Schell, C., and Kornacker, P. (1990). Linear barriers to arthropod movements in the landscape. *Biological Conservation* **54**, 209–22.

Majer, J. D. (1983). Ants: bio-indicators of minesite rehabilitation, land-use and land conservation. *Environmental Management* **7**, 375–83.

Majer, J. D. (1985). Recolonisation by ants of rehabilitated mineral sand mines, on North Stradbroke Island, Queensland, with particular reference to seed removal. *Australian Journal of Ecology* **10**, 31–48.

Malausa, J. C. and Drescher, J. (1991). The project to rescue the Italian ground beetle *Chrysocarabus olympiae*. *International Zoo Yearbook* **30**, 75–79.

Manton, S. M. (1977). *The Arthropoda*. Clarendon Press, Oxford.

Marshall, A. G. (1982). The butterfly industry of Taiwan. *Antenna* **6**, 203–4.

Marshall, A. J. and Williams, W. D. (ed.) (1982). *Textbook of zoology. Invertebrates*. Macmillan, London.

Master, L. R. (1991). Assessing threats and setting priorities for conservation. *Conservation Biology* **5**, 559–63.

Mattoni, R. H. T. (1992). The endangered El Segundo blue butterfly. *Journal of Research on the Lepidoptera* **29**, 277–304.

Mattoni, R. H. T. (1993). The El Segundo Blue, *Euphilotes bernardino allyni* (Shields). **In** *Conservation biology of Lycaenidae* (ed. T. R. New), pp.133–4. IUCN, Gland.

May, R. M. (1988). How many species are there on earth? *Science* **241**, 1441–9.

May, R. M. (1990). How many species? *Philosophical Transactions of the Royal Society of London B*. **330**, 293–304.

Mead, A. R. (1961). *The giant african snail: a problem in economic malacology*. University of Chicago Press, Chicago.

Mead, A. R. (1979). Economic malacology, with particular reference to *Achatina fullica*. **In** *The Pulmonates*, Vol. 2B (ed. V. Fretter, and J. Peake), Academic Press, London.

Meads, M. J. (1987*a*). The giant weta (*Deinacrida heteracantha*) at Mahoenui, King Country: present status and strategy for saving the species. *Reports of the Ecological Division, DSIR* **7**, 1–7.

Meads, M. J. (1987*b*). The giant weta (*Deinacrida parva*) at Puhi, Kaikowa: present status and strategy for saving the species. *Reports of the Ecological Division, DSIR* **8**, 1–8.

Meads, M. J. (1988). A brief review of the current status of the giant wetas and a strategy for their conservation. *Reports of the Ecological Division, DSIR* **11**, 1–11.

Meads, M. (1990). *The weta book. A guide to the identification of wetas*. DSIR Land Resources, Lower Hutt.

Meads, M. J. and Moller, H. (1978). Introduction of giant wetas *Deinacrida rugosa* to Maud Island and observations of tree wetas, paryphantids and other invertebrates. Ecological Division DSIR, unpublished report.

Meads, M. J., Walker, K. J., and Elliott, G. P. (1984). Status, conservation, and management of the land snails of the genus *Powelliphanta*

(Mollusca: Pulmonata). *New Zealand Journal of Zoology* **11**, 277–306.

Meglitsch, P. A. and Schram, F. R. (1991). *Invertebrate Zoology*, (3rd edn). Oxford University Press, New York.

Minkin, B. I. (1990). Leeches in modern medicine. *Carolina Tips* **53**, (2), 5–6.

Mochida, O. (1991). Spread of freshwater *Pomacea* snails (Pilidae, Mollusca) from Argentina to Asia. *Micronesica* (Suppl. **3**), 51–62.

Moller, H. and Tilley, J. A. V. (1989). Beech honeydew: seasonal variation and use by wasps, honeybees and other insects. *New Zealand Journal of Zoology* **16**, 289–302.

Mooney, H. A. and Drake, J. A. (ed.) (1986). *Ecology of biological invasions of North America and Hawaii*. Springer Verlag, New York.

Moore, N. W. (1969). Experience with pesticides and the theory of conservation. *Biological Conservation* **1**, 201–7.

Moore, N. W. (1987). *The bird of time*. Cambridge University Press, Cambridge.

Moreno, C. A., Sutherland, J. P., and Jara, H. F. (1984). Man as a predator in the intertidal zone of southern Chile. *Oikos* **42**, 155–60.

Moriarty, F. (1983). *Ecotoxicology*. Academic Press, New York.

Morris, M. G. (1971). The management of grassland for the conservation of invertebrate animals. **In** *The scientific management of animal and plant communities for conservation* (ed. E. Duffey, and A. S. Watt), pp. 527–52. Blackwell, Oxford.

Morris, M. G. (1979). Grassland management and invertebrate animals – a selective review. *Scientific Proceedings of the Royal Dublin Society, A* **6**, 129–39.

Morris, M. G. and Webb, N. R. (1987). The importance of field margins for the conservation of insects. *BCPC Monograph* **35**, 53–65.

Morrissy, N. M., Hall, N. and Kaputi, N. (1986). *A bioeconomic model for semi-intensive grow-out of marron* (Cherax tenuimanus). Fisheries Management Discussion Paper No. **2**, pp. 93–139, Fisheries Department of Western Australia, Perth.

Morton, A. C. (1983). Butterfly conservation – the need for a captive breeding institute. *Biological Conservation* **25**, 19–33.

Morton, A. C. (1991). Captive breeding of butterflies and moths. I. Advances in equipment and techniques. II. Conserving genetic variation and managing biodiversity. *International Zoo Yearbook* **30**, 80–9, 89–97.

Morton, S. R. and James, C. D. (1988). The diversity and abundance of lizards in arid Australia: a new hypothesis. *American Naturalist* **132**, 237–56.

Mound, L. A. and Gaston, K. J. (1994). Conservation and systematics – the agony and the ecstasy. In *Perspectives on insect conservation*. (ed. K. J. Gaston, T. R. New, and M. J. Samways), pp.185–95. Intercept, Andover, Hants.

Munguira, M. L. and Thomas, J. A. (1992). Use of road verges by butterfly and burnet populations, and the effects of roads on adult dispersal and mortality. *Journal of Applied Ecology* **29**, 316–29.

Munthali, S. M. and Mughogho, D. E. C. (1992). Economic incentives for conservation: bee-keeping and Saturniidae caterpillar utilisation by rural communities. *Biodiversity and Conservation* **1**, 143–54.

Murphy, D. D. (1989). Are we studying our endangered butterflies to death? *Journal of Research on the Lepidoptera* **26**, 236–9.

Murphy, D. D., Freas, K. E., and Weiss, S. B. (1990). An environment-metapopulation approach to population viability analysis for a threatened invertebrate. *Conservation Biology* **4**, 41–51.

Murray, J. and Clarke, B. (1980). The genus *Partula* on Moorea: speciation in progress. *Proceedings of the Royal Society of London B* **211**, 83–117.

Murray, J., Murray, E., Johnson, M. S., and Clarke, B. (1988). The extinction of *Partula* on Moorea. *Pacific Science* **42** 150–3.

Murray, N. D. (1985). Rates of change in introduced organisms. In *Proceedings of the VI International Symposium on the Biological Control of Weeds, Vancouver* (ed. E. S. Delfosse), pp. 191–9. Agriculture Canada, Ottawa.

Myers, K. (1986). Introduced vertebrates in Australia, with emphasis on the mammals. In *Ecology of biological invasions: an Australian perspective* (ed. R. H. Groves, and J. J. Burdon), pp. 120–36. Australian Academy of Science, Canberra.

Myers, N. (1983). *A wealth of wild species*. Westview Press, Boulder, Colorado.

Nafus, D. M. (1993). Extinction, biological control and insect conservation on islands. In *Perspectives on insect conservation* (ed. K. J. Gaston, T. R. New, and M. J. Samways,) pp. 139–54. Intercept, Andover, Hants.

National Research Council (1983). *Butterfly farming in Papua New Guinea*. National Academy Press, Washington, DC.

National Water Council (1981). *River quality: the 1980 survey and future outlook*. National Water Council, London.

Naylor, E. (1965). Effects of heated effluents upon marine and estuarine organisms. *Advances in Marine Biology* **3**, 63–103.

Neel, J. K. (1953). Certain limnological features of a polluted irrigation stream. *Transactions of the American Microscopical Society* **72**, 119–35.

Nentwig, W. (1988). Augmentation of beneficial arthropods by strip management. 1. Succession of predacious arthropods and long-term change in the ratio of phytophagous and predacious arthropods in a meadow. *Oecologia* **76**, 597–606.

Neves, R. J. and Odom, M. C. (1989). Muskrat predation on endangered freshwater mussels in Virginia. *Journal of Wildlife Management* **53**, 934–41.

New, T. R. (1984). *Insect conservation: an Australian perspective*. W. Junk, Dordrecht.

New, T. R. (1987). Insect conservation in Australia: towards rational ecological priorities. In *The role of invertebrates in conservation and biological survey* (ed. J. D. Majer) pp. 5–20. Department of Conservation and Land Management, Perth.

New, T. R. (1991*a*). *Butterfly conservation*. Oxford University Press, Melbourne.

New, T. R. (1991*b*). The doctor's dilemma: or ideals, attitudes and practicality in insect conservation. *Journal of the Australian Entomological Society* **30**, 91–108.

New, T. R. (ed.) (1993*a*). *Conservation biology of the Lycaenidae (Lepidoptera)*. IUCN, Gland.

New, T. R. (1993*b*). Angels on a pin: dimensions of the crisis in invertebrate conservation. *American Zoologist* **33**, 623–30.

New, T. R. (1994*a*). Effects of exotic species on Australian native insects In *Perspectives on insect conservation*. (ed. K. J. Gaston, T. R. New, and M. J. Samways), pp. 155–69. Intercept, Andover, Hants.

New, T. R. (1994*b*). *Exotic insects in Australia*. Gleneagles Press, Adelaide.

New, T. R. (1995). Onychophora in invertebrate conservation: priorities, practice and prospects.

Zoological Journal of the Linnean Society (in press).

New, T. R. and Collins, N. M. (1991). *Swallowtail butterflies. An action plan for their conservation.* IUCN, Gland.

Newsom, L. D. (1967). Consequences of insecticide use on non-target organisms. *Annual Review of Entomology* **12**, 257–86.

Nielsen, L. J. and Orth, D. J. (1988). The hellgrammite-crayfish bait industry of the New River and its tributaries, West Virginia. *North American Journal of Fisheries Management* **8**, 317–24.

Niven, B. S. (1988). Logical synthesis of an animal's environment: sponges to non-human primates. V. The Cane Toad, *Bufo marinus*. *Australian Journal of Zoology* **36**, 169–94.

Norton, B. G. (ed.) (1986). *The preservation of species*. Princeton University Press, Princeton, NJ.

Noss, R. F. (1990). Indicators for monitoring biodiversity: a hierarchical approach. *Conservation Biology* **4**, 355–64.

Notman, P. (1989). A review of invertebrate poisoning by compound 1080. *New Zealand Entomologist* **12**, 67–71.

Olney, P. J. S. and Ellis, P. (eds.) (1991). *1990 International Zoo Year Book*. Zoological Society of London, London.

Opler, P. A. (1991). North American problems and perspectives in insect conservation. In *The conservation of insects and their habitats* (ed. N. M. Collins, and J. A. Thomas), pp.9–32. Academic Press, London.

Orsak, L. (1993). Killing butterflies to save butterflies: a tool for tropical forest conservation in Papua New Guinea. *News Bulletin of the Lepidopterists' Society* May/June 1993, 71–80.

Owen, D. F. (1971). *Tropical butterflies*. Clarendon Press, Oxford.

Paine, R. T. (1974). Intertidal community structure. Experimental studies on the relationship between a dominant competitor and its principal predator. *Oecologia* **15**, 93–120.

Palmer, M. A., Bely, A. E., and Berg, K. E. (1992). Response of invertebrates to lotic disturbance: a test of the hyporheic refuge hypothesis. *Oecologia* **89** 182–94.

Parsons, M. J. (1978). *Farming manual: insect farming and trading agency*. Division of Wildlife, Papua New Guinea.

Parsons, M. J. (1983). A conservation study of the birdwing butterflies, Ornithoptera and Troides (Lepidoptera: Papilionidae) in Papua New Guinea. Report to the Department of Primary Industry, Papua New Guinea.

Parsons, M. (1984). The biology and conservation of *Ornithoptera alexandrae*. In *The biology of butterflies* (ed. R. I. Vane-Wright, and P. R. Ackery), pp. 327–31. Academic Press, London.

Parsons, P. A. (1983). *The evolutionary biology of colonising species*. Cambridge University Press, Cambridge.

Parsons, P. A. (1989). Conservation and global warming: a problem in biological adaptation to stress. *Ambio* **18**, 322–5.

Parsons, P. A. (1990). Fluctuating asymmetry: an epigenetic measure of stress. *Biological Reviews* **65**, 131–45.

Pearce-Kelly, P., Clarke, D., Robertson, M., and Andrews, C. (1991). The display, culture and conservation of invertebrates at London Zoo. *International Zoo Yearbook* **30**, 21–30.

Pearman, G. I. (ed.) (1988). *Greenhouse. Planning for climatic change*. CSIRO, Melbourne.

Pearson, D. L. and Cassola, F. (1992). World-wide species richness patterns of tiger beetles (Coleoptera: Cicindelidae): indicator taxon for biodiversity and conservation studies. *Conservation Biology* **6**, 376–91.

Pellew, R. A. (1991). Data management for conservation. In *The scientific management of temperate communities for conservation* (ed. I. F. Spellerberg, F. B. Goldsmith, and M. G. Morris), pp. 505–22. Blackwell Scientific Publications, Oxford.

Peters, R. L. and Darling, J. D. S. (1985). The greenhouse effect and nature reserves. *BioScience* **35**, 707–17.

Petersen, L. B.-M. and Petersen, R. C. (1983). Anomalies in hydropsychid capture nets from polluted streams. *Freshwater Biology* **13**, 185–91.

Pettigrove, V. (1989). Larval mouthpart deformities in *Procladius paludicola* Skuse (Diptera: Chironomidae) from the Murray and Darling Rivers, Australia. *Hydrobiologica* **179**, 111–17.

Pianka, E. R. (1989). Desert lizard diversity; additional comments and some data. *American Naturalist* **134**, 344–64.

Platnick, N. I. (1991). Patterns of biodiversity: tropical vs temperate. *Journal of Natural History* **25**, 1083–8.

Polhemus, D. A. (1993). Conservation of aquatic insects: worldwide crisis or localised threats? *American Zoologist* **33**, 588–98.

Pollard, E. (1975). Aspects of the ecology of *Helix pomatia* L. *Journal of Animal Ecology* **44**, 305–29.

Pollard, E., Hooper, M. D., and Moore, N. W. (1974). *Hedges*. Collins, London.

Ponder, W. F. and Clark, G. A. (1990). A radiation of hydrobiid snails in threatened artesian springs in western Queensland. *Records of the Australian Museum* **42**, 301–63.

Poore, G. C. B. and Wilson, G. D. F. (1993). Marine species richness. *Nature* **361**, 597–8.

Port, G. R. and Spencer, H. J. (1987). Effects of roadside conditions on some Auchenorrhyncha. **In** *Proceedings of the 6th Auchenorrhyncha meeting* (ed. C. Vidano, and A. Arzone), pp. 181–7. Consiglio Nazionale delle Richerche-IPRA, Rome.

Port, G. R. and Thompson, J. R. (1980). Outbreaks of insect herbivores on plants along motorways in the United Kingdom. *Journal of Applied Ecology* **17**, 649–56.

Prchal, S. J. (1991). Sonoran Arthropod Studies, Inc.: a new concept in environmental education. *International Zoo Yearbook* **30**, 40–5.

Pyle, R. M. (1976). Conservation of Lepidoptera in the United States. *Biological Conservation* **9**, 55–75.

Pyle, R., Bentzien, M. and Opler, P. (1981). Insect conservation. *Annual Review of Entomology* **26**, 233–58.

Quinn, J. F., Wing, S. R., and Botsford, L. W. (1993). Harvest refugia in marine invertebrate fisheries; models and applications to the red sea urchin, *Strongylocentrotus franciscanus*. *American Zoologist* **33**, 537–50.

Quinn, J. M., Davies-Colley, R. J., Hickey, C. W., Vickers, M. L. and Ryan, P. A. (1992). Effects of clay discharges on streams. 2. Benthic invertebrates. *Hydrobiologia* **248**, 235–47.

Rabinowitz, D. (1981). Seven forms of rarity. In *The biological aspects of rare plant conservation* (ed. H. Synge), pp. 205–17. Wiley, New York.

Rabinowitz, D., Cairns, S., and Dillon, T. (1986). Seven forms of rarity and their frequency in the flora of the British Isles. **In** *Conservation Biology; the science of scarcity and diversity*. (ed. M. E. Soulé), pp. 182–204. Sinauer, Sunderland, Mass.

Racek, A. A. (1969). The freshwater sponges of Australia. *Australian Journal of Marine and Freshwater Research* **20**, 267–310.

Raddum, G. G. and Fjellheim, A. (1984). Acidification and early warning organisms in freshwater in western Norway. *Verhandlungen der Internaionalen Vereinigung fur Theoretische und Angewandte Limnologie*. **22**, 1973–80.

Ramsay, G. W., Meads, M. J., Sherley, G. H., and Gibbs, G. W. (1988). Wildlife Research Liaison Group – research on terrestrial insects of New Zealand. *Wildlife Research Liaison Group Research Reviews* **10**, 1–49.

Rands, M. R. W. (1985). Pesticide use on cereals and the survival of grey partridge chicks: a field experiment. *Journal of Applied Ecology* **22**, 49–54.

Rands, M. R. W. and Sotherton, N. W. (1986). Pesticide use on cereal crops and changes in the abundance of butterflies on arable farmland. *Biological Conservation* **36**, 71–82.

Ratcliffe, D. A. (ed.) (1977). *A nature conservation review*, Vols 1 and 2. Cambridge University Press, Cambridge.

Ravetto, P., Canaglia, D., Colombo, V., and Peila, V. (1987). [Proposal for utilisation of the honey bee as an efficient indicator of radioactive contamination.] *l'Apicoltore Moderno* **78**, 187–91.

Regan, D. H. (1986). Duties of preservation. **In** *The preservation of species* (ed. B. G. Norton), pp. 195–220. Princeton University Press, Princeton, New Jersey.

Reice, S. R., Wissmar, R. C., and Naiman, R. J. (1990). Disturbance regimes resilience, and recovery of animal communities and habitats in lotic ecosystems. *Environmental Management* **14**, 647–59.

Resh, V. H. and Sorg, K. L. (1983). Distribution of the Wilbur Springs shore bug (Hemiptera: Saldidae): predicting occurrence using water chemistry parameters. *Environmental Entomology* **12**, 1628–35.

Resh, V. H., *et al.* (1988). The role of disturbance in stream ecology. *Journal of the North American Benthological Society* **7**, 433–55.

Reynolds, J. D. (1988). Crayfish extinctions and crayfish plague in central Ireland. *Biological Conservation* **45**, 279–85.

Rhett, R.G., Simmers, J. W., and Lee, C. R. (1988). *Eisenia foetida* used as a biomonitoring tool to predict the potential bioaccumulation of contaminants from contaminated dredged material. In *Earthworms in waste and environmental management* (ed. C. A. Edwards, and E. F. Neuhauser), pp. 321–8. SPC, The Hague.

Robertson, P. A., Woodburn, M. I. A., and Hill, D. A. (1988). The effects of woodland management for pheasants on the abundance of butterflies in Dorset, England. *Biological Conservation* **45**, 159–67.

Robinson, M. H. (1991). Invertebrates: exhibiting the silent majority. *International Zoo Yearbook* **30**, 1–7.

Root, M. (1990). Biological monitors of pollution. *BioScience* **40**, 83–8.

Rosenberg, D. M., Danks, H. V., and Lehmkuhl, D. M. (1986). Importance of insects in environmental impact assessment. *Environmental Management* **10**, 773–83.

Rounsefell, G. A. (1975). *Ecology, utilization, and management of marine Fisheries*. Mosby, Saint Louis.

Rózsa, L. (1992). Endangered parasite species. *International Journal of Parasitology* **22**, 265–6.

Ruggieri, G. D. (1976). Drugs from the sea. *Science* **194**, 491–7.

Rushton, S. P., Eyre, M. D., and Luff, M. L. (1989). The effects of management on the occurrence of some carabid species in grassland. In *The role of ground beetles in ecological and environmental studies* (ed. N. E. Stork), pp. 209–16. Intercept, Andover, Hants.

Russell, R. C., Rajapaks, N., Whelan, P. I., and Langsford, W. H. (1984). Mosquito and other insect introductions to Australia aboard international aircraft, and the monitoring of disinsection procedures. In *Commerce and the spread of pests and diseases*, (ed. M. Laird), pp. 109–41. Praeger, New York.

SA (1994). *Systematics Agenda 2000: Charting the biosphere*. Technical Report.

Sabine, J. R. (1988). Vermiculture: bring on the future. In *Earthworms in waste and environmental management* (ed. C. A. Edwards, and E. F. Neuhauser), pp. 3–7. SPC Academic Publishing, The Hague.

Saila, S. B. (1983). Importance and assessment of discards in commercial fisheries. FAO Fisheries Circular, No. 765, 1–62.

Salm, R. V. (1984). Ecological boundaries for coral-reef reserves: principles and guidelines. *Environmental Conservation* **11**, 209–15.

Samson, P. R. (1993). Illidge's ant-blue, *Acrodipsas illidgei* (Waterhouse and Lyell). In *Conservation biology of the lycaenidae* (Lepidoptera) (ed. T. R. New), pp. 163–5. IUCN, Gland.

Samways, M. J. (1988). Classical biological control and insect conservation: are they compatible? *Environmental Conservation* **15**, 349–54, 348.

Samways, M. J. (1989a). Insect conservation and landscape ecology: a case-history of bush crickets (Tettigoniidae) in southern France. *Environmental Conservation* **16**, 217–26.

Samways, M. J. (1989b). Insect conservation and the disturbance landscape. *Agriculture, Ecosystems and Environment* **27**, 183–94.

Samways, M. J. (1994a). A spatial and process subregional framework for insect and biodiversity conservation research and management. In *Perspectives on Insect Conservation* (ed. K. J. Gaston, T. R. New, and M. J. Samways), pp. 1–27. Intercept, Andover, Hants.

Samways, M. J. (1994b). *Insect conservation biology*. Chapman & Hall, London.

Sant, G. J. and New, T. R. (1988). *The biology and conservation of* Hemiphlebia mirabilis Selys *(Odonata, Hemiphlebiidae) in southern Victoria*. Technical Report Series No. 82, Arthur Rylah Institute for Environmental Research, Melbourne.

SASI (Sonoran Arthropod Studies, Inc.) (1992). *Annual Report 1991*.

Sawyer, R. T. (1974). Leeches (Annelida: Hirundinea). In *Pollution ecology of freshwater invertebrates* (ed. C. W. Hart, and S. L. H. Fuller), pp. 81–142. Academic Press, New York.

Sawyer, R. T. (1976). The medicinal leech, *Hirudo medicinalis* L., an endangered species. *Proceedings of the 1st South Carolina Endangered Species Symposium*.

Scholtz, C. H. and Chown, S. L. (1994). Insect conservation and extensive agriculture: the savanna of southern Africa. In *Perspectives on insect conservation*. (ed. K. J. Gaston, T. R. New, and M. J. Samways), pp. 75–95. Intercept, Andover, Harts.

Schoon, N. (1992). Zoo backs effort to save the snail. *The Independent*, 28 August, p. 7.

Scott, I. A. W. and Rowell, D. M. (1991). Population biology of *Euperipatoides leuckartii* (Onychophora: Peripatopsidae). *Australian Journal of Zoology* **39**, 499–508.

Selander, R. K. and Ochman, H. (1983). The

genetic structure of populations as illustrated by molluscs. Isozymes. *Current Topics in Biological and Medical Research (Genetics and Evolution).* **10**, 93–123.

Sheppard, D. (1991). Site survey methods. In *Habitat conservation for insects – a neglected green issue* (ed. R. Fry, and D. Lonsdale), pp. 205–8. Amateur Entomologists' Society, Middlesex.

Sherley, G. H. and Hayes, L. M. (1993). The conservation of a giant weta (*Deinacrida* n. sp. Orthoptera: Stenopelmatidae) at Mahoenui, King Country: habitat use, and other aspects of its ecology. *New Zealand Entomologist* **16**, 55–68.

Shirt, D. B. (ed.) (1987). *British red data books: 2. Insects.* Nature Conservancy Council, Peterborough.

Shreeve, T. G. and Mason, C. F. (1980). The number of butterfly species in woodlands. *Oecologia* **45**, 414–18.

Simberloff, D. S. (1986). Introduced insects: a biogeographic and systematic perspective. In *Ecology of Biological Invasions of North America and Hawaii.* (ed. H. A. Mooney, and J. A. Drake), pp. 3–26. Springer-Verlag, New York.

Simpson, K. W. (1980). Abnormalities in the tracheal gills of aquatic insects collected from streams receiving chlorinated or crude oil wastes. *Freshwater Biology* **10**, 581–3.

Sinderman, N. J. C. (1986). Strategies for reducing risks from introductions of aquatic organisms: a marine perspective. *Fisheries* **11**, 10–15.

Smith, B. C. and Kershaw, R. A. (1981). *Tasmanian land and freshwater molluscs.* Fauna of Tasmania Handbook No. 5. University of Tasmania, Hobart.

Smith, B. C. and Peterson, J. A. (1982). Studies on the Giant Gippsland Earthworm, *Megascolides australis* McCoy 1878. *Victorian Naturalist* **99**, 164–73.

Smith, R. H. and Holloway, G. J. (1989). Population genetics and insect introductions. *Entomologist* **108**, 14–27.

Solbrig, O. T. (1991). *From genes to ecosystems: a research agenda for biodiversity.* IUBS Monograph Series.

Solem, A. (1990). How many Hawaiian land snail species are left? and what we can do for them. *Occasional Papers of the Bishop Museum, Honolulu* **30**, 27–40.

Sotherton, N. W., Boatman, N. D., and Rands,

M. R. W. (1989). The conservation headland experiment in cereal ecosystems. *Entomologist* **108**, 135–43.

Soulé, M. E. (1980). Thresholds for survival: maintaining fitness and evolutionary potential. In *Conservation biology: an evolutionary–ecological perspective.* (ed. M. E. Soulé, and B. A. Wilcox), pp. 151–69. Sinauer, Associates, Sunderland, Massachusetts.

Soulé, M. and Baker, B. (1968). Phenetics of natural populations. IV. the population asymmetry parameter in the butterfly *Coenonympha tullia. Heredity* **23**, 611–14.

Soulé, M. E. and Kohm, K. A. (1989). *Research priorities for conservation biology.* Society for Conservation Biology and Island Press, Washington, DC.

Southwood, T. R. E. (1977). Habitat, the templet for ecological strategies? *Journal of Animal Ecology* **46**, 337–66.

Southwood, T. R. E. (1978). *Ecological Methods,* (2nd edn). Chapman & Hall, London.

Speight, M. C. D. (1986*a*). Attitudes to insects and insect conservation. *Proceedings of the 3rd European Congress of Entomology* Vol. **13**, pp. 369–85.

Speight, M. C. D. (1986*b*). Criteria for the selection of insects to be used as bioindicators in nature conservation research. *Proceedings of the 3rd European Congress of Entomology,* Vol. **3**, pp. 485–8.

Speight, M. C. D. (1989). Saproxylic invertebrates and their conservation. *Nature and Environment Series* **42**, 1–82. Council of Europe, Strasbourg.

Spellerberg, I. F. (1993). *Monitoring ecological change.* Cambridge University Press, Cambridge.

Stanford, A. J. and Koznecous, J. (1988). Aquaculture of the yabbie, *Cherax destructor* Clark (Decapoda: Parastacidae): an economic evaluation. *Aquaculture Fisheries Management* **19**, 325–40.

Stork, N. E. (1988). Insect diversity: facts, fiction and speculation. *Biological Journal of the Linnean Society* **35**, 321–37.

Stork, N. E. (ed.) (1990). *The role of ground beetles in ecological and environmental studies.* Intercept, Andover.

Stork, N. E. (1991). The composition of the arthropod fauna of Bornean lowland rain forest trees. *Journal of Tropical Ecology* **7**, 161–80.

Strong, L. (1992). Avermectins: a review of their

impact on insects of cattle dung. *Bulletin of Entomological Research* **82**, 265–74.

Stroot, P. (1987). An attempt to evaluate the state of the caddis fly fauna of Belgium. *Proceedings of the 5th international Symposium on Trichoptera*, pp. 79–93. (W. Junk, The Hague).

Stroot, P. and Depiereux, E. (1989). Proposition d'une Méthodologie pour Etablir des 'Listes Rouges' d'Invertébrés menacés. *Biological Conservation* **48**, 163–79.

Sundaram, K. M.S., Holmes, S. B., Kreutzweiser, D. P., Sundaram, A., and Kingsbury, P. D. (1991). Environmental persistence and impact of Diflubenzuron in a forest aquatic environment following aerial application. *Archives of Environmental Contamination and Toxicology* **20**, 313–24.

Sutton, S. L. and Collins, N. M. (1991). Insects and tropical forest conservation. In *The conservation of insects and their habitats* (ed. N. M. Collins, and J. A. Thomas), pp. 405–24. Academic Press, London.

Swift, M. J., Heal, O. W., and Anderson, J. M. (1979). *Decomposition in terrestrial ecosystems*. Blackwell Scientific Publications, Oxford.

Tait, N. N., Stutchbury, R. J., and Briscoe, D. A. (1990). Review of the discovery and identification of Onychophora in Australia. *Proceedings of the Linnean Society of New South Wales* **112**, 153–71.

Taugbol, T., Skurdal, J., and Håstein, T. (1993). Crayfish plague and management strategies in Norway. *Biological Conservation* **63**, 75–82.

Taylor, R. W. (1976). Submission to the inquiry into the impact on the Australian environment of the current woodchip industry programme. In Australian Senate Official Hansard Report (reference: Woodchip Inquiry). pp. 3724–31.

Taylor, R. W. (1983). Descriptive taxonomy: past, present and future. In *Australian systematic entomology: a bicentenary perspective* (ed. E. Highley, and R. W. Taylor), pp. 93–134. CSIRO, Melbourne.

Taylor, S. (1993). Practical ecosystem management for plants and animals. *Endangered Species Update* **10**, 26–9.

Templeton, A. R. (1991). Genetics and conservation biology. In *Species conservation: a population biological approach* (ed. A. Seitz, and V. Loeschcke), pp. 15–29. Birkhauser Verlag, Basel.

Thomas, C. D., Moller, H., Plunkett, G. M., and Harris, R. J. (1990). The prevalence of introduced *Vespula vulgaris* wasps in a New Zealand beech forest community. *New Zealand Journal of Ecology* **13**, 63–72.

Thomas, J. A. (1983). The ecology and conservation of *Lysandra bellargus* (Lepidoptera: Lycaenidae) in Britain. *Journal of Applied Ecology* **20**, 59–83.

Thomas, J. A. (1984). The conservation of butterflies in temperate countries: past efforts and lessons for the future. In *The biology of butterflies* (ed. R. I. Vane-Wright, and P. R. Ackery), pp. 333–53. Academic Press, London.

Thomas, J. A. (1989). Ecological lessons from the re-introduction of Lepidoptera. *Entomologist* **108**, 56–68.

Thomas, J. A. (1991). Rare species conservation: case studies of European butterflies. In *The scientific management of temperate communities for conservation*. (ed. I. F. Spellerberg, M. G. Morris, and F. B. Goldsmith), pp. 141–97. Blackwell Scientific Publications, Oxford.

Thomas, M. B., Wratten, S. D., and Sotherton, N. W. (1991). Creation of 'island' habitats in farmland to manipulate populations of beneficial arthropods: predator densities and emigration. *Journal of Applied Ecology* **28**, 906–17.

Thomas, M. B., Wratten, S. D., and Sotherton, N. W. (1992). Creation of 'island' habitats in farmland to manipulate populations of beneficial arthropods: predator densities and species composition. *Journal of Applied Ecology* **29**, 524–31.

Thorson, G., (1971). *Life in the sea*. McGraw-Hill, New York.

Tilzey, R. J. D. (1977). Key factors in the establishment and success of trout in Australia. *Proceedings of the Ecological Society of Australia* **10**, 97–105.

Tonge, S. and Bloxam, Q. (1991). A review of the captive-breeding programme for Polynesian tree snails. *International Zoo Yearbook* **30**, 51–9.

Tranvik, L. and Eijsackers, H. (1989). On the advantage of *Folsomia fimetarioides* over *Isotomiella minor* (Collembola) in a metal polluted soil. *Oecologia* **80**, 195–200.

Tranvik, L., Bengtsson, G., and Rundgren, S. (1993). Relative abundance and resistance traits of two Collembola species under metal stress. *Journal of Applied Ecology* **30**, 43–52.

Trueman, J. W. H., Hoye, G. A., Hawking,

J. H., Watson, J. A. L., and New, T. R. (1992). *Hemiphlebia mirabilis* Selys: new localities in Australia and perspectives on conservation (Zygoptera: Hemiphlebiidae). *Odonatologica* **21**, 367–74.

Uglem, G. L., Lewis, M. C., and Short, T. M. (1990). Contributions to the life history of *Proterometra dickermani* (Digenea: Azygiidae). *Journal of Parasitology* **76**, 447–50.

Underwood, A. J. and Kennelly, S. J. (1990). Pilot studies for designs of surveys of human disturbance of intertidal habitats in New South Wales. In *The effects of fishing* (ed. W. Craik, J. Glaister, and I. Poiner), pp. 165–73. CSIRO, Melbourne.

Unestam, T. (1973). Significance of diseases in freshwater crayfish. *Freshwater Crayfish* **2**, 136–50.

Usher, M. B. (1986). Wildlife conservation evaluation: attitudes, criteria and values. In *Wildlife conservation evaluation* (ed. M. B. Usher), pp. 4–44. Chapman & Hall, London.

Usher, M. B. (1989). Scientific aspects of nature conservation in the United Kingdom. *Journal of Applied Ecology* **26**, 813–24.

Usher, M. B. and Edwards, M. (1986). The selection of conservation areas in Antarctica: an example using the arthropod fauna of antarctic islands. *Environmental Conservation* **13**, 115–22.

Usher, M. B. and Jefferson, R. G. (1991). Creating new and successional habitats for arthropods. In *The conservation of insects and their habitats* (ed. N. M. Collins, and J. A. Thomas), pp. 263–91 Academic Press, London.

van Praagh, B. (1992). The biology and conservation of the Giant Gippsland Earthworm, *Megascolides australis* McCoy 1878. *Soil Biology and Biochemistry* **24**, 1363–7.

van Praagh, B. D. (1994). The biology and conservation of *Megascolides australis* McCoy 1878. Ph. D. Thesis, La Trobe University, Melbourne.

van Praagh, B. D., Yen, A. L., and Lillywhite, P. K. (1989). Further information on the Giant Gippsland Earthworm, *Megascolides australis* (McCoy 1878). *Victorian Naturalist* **106**, 197–201.

van Tol, J. and Verdonk, M. J. (1988). *The protection of dragonflies (Odonata) and their biotopes*. Nature and environment series No. 38, Council of Europe, Strasbourg.

Vane-Wright, R. I. (1991). Why not eat insects? *Bulletin of Entomological Research* **81**, 1–4.

Vane-Wright, R. I. (1994). Systematics and the conservation of biodiversity: global, national and local perspectives. In *Perspectives on insect conservation*. (ed. K. J. Gaston, T. R. New, and M. J. Samways), pp. 197–211. Intercept, Andover, Hants.

Vane-Wright, R. I., Humphries, C. J., and Williams, P. H. (1991). What to protect? Systematics and the agony of choice. *Biological Conservation* **55**, 235–54.

Vermeij, G. J. (1978). *Biogeography and adaptation. Patterns of marine life*. Harvard University Press, Cambridge, Mass.

Vinson, S. B. and Greenberg, L. (1986). The biology, physiology and ecology of imported fire ants. In *Economic impact and control of social insects* (ed. S. B. Vinson), pp. 193–226. Praeger, New York.

Walker, B. H. (1992). Biodiversity and ecological redundancy. *Conservation Biology* **6**, 18–23.

Walter, H. and Breckle, S.-W. (1989). *Ecological systems of the Geobiosphere. 3.Temperate and polar zonobiomes of northern Eurasia*. Springer-Verlag, Berlin.

Wapshere, A. J. (1974). A strategy for evaluating the safety of organisms for biological weed control. *Annals of Applied Biology* **77**, 201–11.

Warren, M. S. (1985). The influence of shade on butterfly numbers in woodland rides, with special reference to the wood white, *Leptidea sinapis*. *Biological Conservation*, **33**, 147–64.

Warwick, W. F., Fitchko, J., McKee, P. M., Hart, D. R., and Burt, A. J. (1987). The incidence of deformities in *Chironomus* spp. from Port Hope Harbour, Lake Ontario. *Journal of Great Lakes Research* **13**, 88–92.

Wassenberg, T. J. and Hill, B. J. (1990). Partitioning of material discarded from prawn trawlers in Moreton Bay. In *The effects of fishing* (ed. W. Craik, J. Glaister, and I. Poiner), pp. 27–36. CSIRO, Melbourne.

Weatherley, N. S. and Ormerod, S. J. (1990). The constancy of invertebrate assemblages in softwater streams: implications for the prediction and detection of environmental change. *Journal of Applied Ecology* **27**, 952–64.

Weatherley, N. S., Lloyd, E. C., Rundle, S. D., and Ormerod, S. J. (1993). Management of conifer plantations for the conservation of stream macroinvertebrates. *Biological Conservation* **63**, 171–6.

Webb, N. R. (1989). Studies on the invertebrate fauna of fragmented heathland in Dorset, UK, and the implications for conservation. *Biological Conservation* **47**, 153–65.

Webb, N. R. and Hopkins, P. J. (1984). Invertebrate diversity on fragmented *Calluna* heathland. *Journal of Applied Ecology* **21**, 921–33.

Webb, N. R., Clarke, R. T., and Nicholas, J. T. (1984). Invertebrate diversity on fragmented *Calluna* heathland: effects of surrounding vegetation. *Journal of Biogeography* **11**, 41–6.

Welch, J. M. and Pollard, E. (1975). The exploitation of *Helix pomatia* L. *Biological Conservation* **8**, 155–60.

Wells, S. M. and Chatfield, J. E. (1992). *Threatened non-marine molluscs of Europe*. Nature and environment series No. 64, Council of Europe, Strasbourg.

Wells, S. M., Pyle, R. M. and Collins, N. M. (1983). *The IUCN invertebrate red data book*. IUCN, Gland.

Whalley, P. (1989). Principles and outcome of introductions. *Entomologist* **108**, 69–76.

White, E. G. (1987). Ecological time frames and the conservation of grassland insects. *New Zealand Entomologist* **10**, 146–52.

White, E. G. (1991). The changing abundance of moths in a tussock grassland, 1962–1989, and 50– to 70–year trends. *New Zealand Journal of Ecology* **15**, 5–22.

Wieser, W., Dallinger, R., and Busch, G. (1977). The flow of copper through a terrestrial food chain. II. Factors influencing the copper content of isopods. *Oecologia* **30**, 265–72.

Wilcox, B. A. and Murphy, D. D. (1985). Conservation strategy: the effects of fragmentation on extinction. *American Naturalist* **125**, 879–87.

Williams, P. H., Humphries, C. J., and Vane-Wright, R. I. (1991). Measuring biodiversity: taxonomic relatedness for conservation priorities. *Australian Systematic Botany* **4**, 665–79.

Williams, W. D. (1981). The Crustacea of Australian inland waters. In *Ecological Biogeography of Australia* (ed. A. Keast), pp. 1101–38. W. Junk, The Hague.

Wilson, E. O. (1987). The little things that run the world (the importance and conservation of invertebrates). *Conservation Biology* **1**, 344–6.

Wilson, F. (1960). *A review of the biological control of insects and weeds in Australia and Australian New Guine*a. Technical Communication No. 1, Commonwealth Institute of Biological Control, Ottawa.

Winner, R. W., Boesel, M. W., and Farrell, M. P. (1980). Insect community structure as an index of heavy metal pollution in lotic ecosystems. *Canadian Journal of Fisheries and Aquatic Sciences* **37**, 647–55.

Wulfhorst, J. (1991). Effects of thermal and organic pollution on macroinvertebrate communities in the River Schwalm, a Northern Hesse hill stream (FGR). *Verhandlungen der Internationalen Vereinigung fur Theoretische und Angewandte Limnologie.* **24**, 2149–55.

Xu Zaifu (1987). The work of Xishuangbanna Tropical Botanic Garden in conserving the threatened plants of the Yunnan tropics. In *Botanic gardens and the world conservation strategy.* (ed. D. Bramwell, O. Hamann, V. Heywood, and H. Synge), pp. 239–53. Academic Press, London.

Yajima, M. (1991). The Insect Ecological Land at Tama Zoo. *International Zoo Yearbook* **30**, 7–15.

Yeates, G. W. (1994). Modification and qualification of the nematode maturity index. *Pedobiologia* **38**, 97–101.

Yen, A. L. (1987). A preliminary assessment of the correlation between plant, vertebrate and Coleoptera communities in the Victorian mallee. In *The role of invertebrates in conservation and biological survey* (ed. J. D. Majer), pp. 73–88. Department of Conservation and Land Management, Perth.

Yen, A. L. (1994). The role of museums and zoos in influencing public attitudes towards insect conservation. In *Perspectives on Insect Conservation.* (ed. K. J. Gaston, T. R. New, and M. J. Samways), pp. 213–29. Intercept, Andover, Hants.

Yen, A. L., New, T. R., Van Praagh, B. and Vaughan, P. J. (1990). Invertebrate conservation: three case studies in south-eastern Australia. In *Management and conservation of small populations* (ed. T. W. Clark, and J. H. Seebeck), pp. 207–24. Chicago Zoological Society, Brookfield, Illinois.

Zabarauskas, P. and Indiviglio, F. (1991). The unseen multitude: design and management of invertebrate displays at the New York Bronx Zoo. *International Zoo Yearbook* **30**, 15–20.

Zimmerman, E. C. (1948). *Introduction. Insects of Hawaii*. Vol. I. University of Hawaii Press, Honolulu.

INDEX